WILLIAM HARVEY
A BIBLIOGRAPHY OF HIS WRITINGS

Doctor William
Harvey

A

BIBLIOGRAPHY OF
THE WRITINGS
OF
DR WILLIAM HARVEY
1578-1657

BY

GEOFFREY KEYNES

M.A., M.D., D.LITT.
F.R.C.S., F.R.C.O.G.

Second edition

CAMBRIDGE
AT THE UNIVERSITY PRESS
1953

CAMBRIDGE UNIVERSITY PRESS
Cambridge, New York, Melbourne, Madrid, Cape Town,
Singapore, São Paulo, Delhi, Mexico City

Cambridge University Press
The Edinburgh Building, Cambridge CB2 8RU, UK

Published in the United States of America by Cambridge University Press, New York

www.cambridge.org
Information on this title: www.cambridge.org/9781107623194

First edition 1928
Second edition 1953
First published 1953
First paperback edition 2013

A catalogue record for this publication is available from the British Library

ISBN 978-1-107-62319-4 Paperback

PREFACE

THE first edition of this *Bibliography* was published in 1928, and on that occasion I wrote in the preface:

'The year 1628, in which William Harvey's treatise *De Motu Cordis* was first published, is now recognised as having seen the birth of modern physiological and biological science. Harvey's writings have therefore acquired a historical importance which places them in a class apart from most of the medical and scientific publications of the seventeenth century, and the three hundredth anniversary of the appearance of his first book seems to be the proper occasion for the production of a full record of their purely bibliographical aspect. The full understanding of Harvey's pre-eminence is reflected in the eagerness shewn by medical libraries and private buyers to collect his books, and in the consequent increase of their market value, which has in the last five years been out of proportion even to the general rise in the value of all works of antiquarian interest.

'Harvey published only three books during his life, so that the task of compiling a detailed bibliography of his works has been a fairly simple one. I have deliberately restricted it to the record of his own works because I have felt that an attempt to widen its scope by including an account of the controversies aroused by the announcement of his discoveries would result in the expenditure of much time without any corresponding advantage to scientific or bibliographical knowledge. Harvey's critics, though vociferous enough in their day, have now mostly sunk into an oblivion in which they may be allowed to remain. Harvey himself has given us the cue by his almost complete abstinence from any controversial writings.'

Up to the year 1928 the only important source of bibliographical information about Harvey's works was the list compiled by Mr Charles Perry Fisher, Librarian to the College of Physicians in Philadelphia, and published in 1912. Three hundred copies of my fuller and more complete bibliography were printed in 1928; the edition was sold out in a few weeks, and the book has been difficult to obtain ever since. In the twenty-four years that have passed there have not been any startling discoveries in connexion with Harvey's books, but the few gaps in the record can now be filled, new entries added, and some errors corrected, though I am glad to say that there do not seem to have been many of these.

The first entry of the bibliography was written on 12 February 1927, in Sir William Osler's library, Lady Osler still living then in Oxford at 13 Norham Gardens and being able to give me free access to Sir William's books. Dr W. W. Francis was also there to be characteristically generous and helpful, and he has recently sent me further notes from the Osler Library at McGill University, Montreal, for the present edition. Since 1928 my own collection of Harvey's works has increased, and I have also extended the number of libraries whose holdings are recorded so as to embrace additional lists from the United States of America and, from Sweden, the remarkable assemblage made by Dr Erik Waller for the great medical library which he has now given to the University of Uppsala. The only private collection included in the record besides my own is that of Mr F. C. Pybus. Sir D'Arcy Power's collection was dispersed after his death and Dr LeRoy Crummer's was bequeathed by him to the University of Michigan. In this field, as in so many others, the day of the private collector is over, the supply of these bibliographical rarities having now largely come to an end. A special exception in the record of copies has been made in favour of the first edition of *De Motu Cordis*. In this instance a careful census of all copies surviving in Europe and America had been made by Dr Ernst Weil and he has kindly allowed me to use his material published in *The Library* in 1944 and amplified by notes that he has made since, so that many additional libraries and private collectors' names have been added here. I am greatly indebted to Dr Weil for his help, and to all the other librarians who have supplied me with information and assistance in making the record as complete as possible.

It is to be noticed that Harvey's aloof character is shown by the absence of any record of a presentation copy of *De Motu Cordis*, or of any other of his books. His own library was presumably mostly lost when the College of Physicians was destroyed in the fire of London in 1666, though a few books are said to have been rescued by Dr Christopher Merrett. One of these, still in the library of the Royal College of Physicians, is *Gabrielis*

Falloppii Opera Omnia, Francofurti, f°, 1584. On the title-page is the signature of Harvey's father-in-law, Lancelot Browne, who has made annotations on the fly-leaves and on the margins throughout; there are also numerous marginal notes by Harvey, with his WH monogram, chiefly in the section *De metallis seu fossilibus*, pp. 303–97. A second volume with Harvey's annotations is in the British Museum—Galen's *Opuscula varia*, ed. Goulston, London, 4°, 1640. Another survivor is *De febribus commentarius ex libris aliquot Hippocratis & Galeni*, by Sylvius (Jacques Du Bois), Venice, 1555, which carries annotations by Fabricius ab Aquapendente, and the signatures of Harvey, dated 1621, and of Dr Richard Mead with the bookplate of Georg Kloss, Frankfort. This fabulous book is now in the Bibliotheca Walleriana at Uppsala University.

In 1928 the *Bibliography* was illustrated by a reproduction of the Ditchingham Hall portrait of Harvey. The present edition has the advantage of being able to use as frontispiece a photograph of the earliest portrait of Harvey showing him as he was about the time that he first promulgated his doctrine of the circulation of the blood, and to reproduce also another authentic representation of Harvey in old age which bears a special relation to the first edition of *De Generatione Animalium*. The importance of these two portraits was first established in my monograph on *The Portraiture of William Harvey*, published by the Royal College of Surgeons of England in 1949. The frontispiece is reproduced by courtesy of the owner, Mr Andrew Lloyd, and the President of the Royal College of Physicians, where the picture now hangs.

The first edition of this *Bibliography* was offered in 1928 as part of the commemoration of the tercentenary of the first publication of *De Motu Cordis*. This revised edition has no particular ceremony to perform, but may serve as a humble tribute to Harvey's lasting fame as one of the greatest of all figures in the history of medicine and the biological sciences.

GEOFFREY KEYNES

LONDON
June, 1952

CONTENTS

ILLUSTRATIONS

PLATES

REPRODUCTIONS OF TITLE-PAGES

ABBREVIATIONS

BM   British Museum
BLO   Bodleian Library, Oxford
BML   Boston Medical Library
BWU   Bibliotheca Walleriana, Uppsala University
CCC   Caius College, Cambridge
CPP   College of Physicians, Philadelphia
FMP   Faculté de Médecine, Paris
GUL   Glasgow University Library
LCW   Library of Congress, Washington
LUM   Library of the University of Michigan (Crummer Collection)
LUN   Library of the University of Nebraska
MGU   McGill University, Medical Library
MSL   Medical Society of London
NLS   National Library of Scotland
NYAM   New York Academy of Medicine
OL   Osler Library, McGill University
RCP   Royal College of Physicians, London
RCPE   Royal College of Physicians, Edinburgh
RCS   Royal College of Surgeons of England
RFG   Royal Faculty of Physicians and Surgeons, Glasgow
RSM   Royal Society of Medicine
StBH   Saint Bartholomew's Hospital
SGL   Surgeon-General's Library, Washington
ULC   University Library, Cambridge
ULE   University Library, Edinburgh
WHML   Wellcome Historical Medical Library
YML   Yale Medical Library

An index of the copies recorded will be found at the end
of the volume with the general index.

The numbering of the entries has been left undisturbed
in this edition.

I

DE MOTU CORDIS

'And I remember that when I asked our famous *Harvey*, in the only Discourse I had with him, (which was but a while before he dyed) What were the things that induc'd him to think of a *Circulation of the Blood*? He answer'd me, that when he took notice that the Valves in the Veins of so many several Parts of the Body, were so Plac'd that they gave free passage to the Blood Towards the Heart, but oppos'd the passage of the Venal Blood the Contrary way: He was invited to imagine, that so Provident a Cause as Nature had not so Plac'd so many Valves without Design: and no Design seem'd more probable, than That, since the Blood could not well, because of the interposing Valves, be Sent by the Veins to the Limbs; it should be Sent through the Arteries, and Return through the Veins, whose Valves did not oppose its course that way.' (Robert Boyle, *A Disquisition about the Final Causes of Natural Things*, London, 1688, pp. 157–8.)

BIBLIOGRAPHICAL PREFACE

Harvey's first and greatest work, usually known by its short title of *De Motu Cordis*, was printed in 1628, but his interest in the subject of the circulation of the blood was doubtless initiated by his association with Fabricius ab Aquapendente during his studentship at the University of Padua in the years 1598 to 1602. It was here that he first became acquainted with his teacher's observations on valves in veins, and it was from these that his own observations and experiments ultimately deduced the facts of the circulation of the blood. 'Harvey's common designation of discoverer of the circulation of the blood suggests that he stumbled suddenly and unexpectedly upon the truth. But Harvey, like other scientific investigators, found no short cut to knowledge. His mind had the acuteness necessary to perceive the inconsistencies which had escaped his predecessors, and he possessed the patience and scientific imagination to piece together the innumerable observations that he had made, both while treating his patients and while dissecting and experimenting on animals.'[1] He had arrived at the results of his observations before he had reached the age of forty, and the *Prelectiones* (see no. 52) or notes, which he used for his lectures and demonstrations on the subject at the College of Physicians, are dated 1616, but it was twelve years before he decided to publish his conclusions to the world. The brief treatise in which he embodied his views he called an 'Anatomical Exercise', but it is in reality a great deal more than this, since it is to be regarded as 'the first record of a complete experimental biological investigation, giving a clear and accurate description of the methods employed to recognize the laws governing an important vital process, a knowledge of which had been till then befogged by mistaken conceptions; for in no other book were the actual facts of the circulation so lucidly and comprehensively stated and

---

[1] This passage is taken from the *postscript* to my edition of the English text of *De Motu Cordis*, edited for the Nonesuch Press in 1928. I have repeated it here as it expresses satisfactorily the point I wish to make.

proved by cogent experiments.'[1] Harvey's treatise is, in fact, more important as a demonstration of scientific method in biological research than as an annunciation of the fact of the circulation of the blood. From this beginning has flowed all subsequent biological knowledge in an ever widening stream.

Harvey's *De Motu Cordis* is therefore justly considered to be one of the most fruitful and important books ever published, and surprise has often been expressed that, instead of publishing it in London, he should have entrusted it to a young publisher, William Fitzer, at Frankfort-on-Main. It is true that Frankfort, with its annual book fair, was then a centre of learning and of science very suitable for the publication of a work written in Latin and of international importance. There were, however, other reasons which are likely to have influenced Harvey in his choice. It was suggested by Dr Archibald Malloch[2] in 1929 that Fitzer, who was an Englishman, was chosen by Harvey because of his acquaintance with Robert Fludd (1576–1637), a celebrated London physician and reputed Rosicrucian. More recently the facts have been very fully investigated by Dr Ernst Weil who has published them in an admirable article in *The Library* (1944), xxiv, 142, to which the reader is referred for details. It seems that Fludd from 1617 had his books published by Johann Theodor de Bry of Oppenheim, formerly of Frankfort. In 1619 the firm returned to Frankfort, and in 1626, when de Bry died, the business passed to Fitzer who then took over the publication of books by Fludd. Harvey had many contacts with Fludd beginning in their student days at Padua in 1602, and is exceedingly likely to have been influenced by him in choosing Fitzer as his publisher. Fitzer's career from 1600 to his death in 1671 has been fully elucidated by Dr Weil but does not further concern us here. The choice of Frankfort, however, was certainly unfortunate, since its distance from London made communication difficult, and it seems to be unlikely that Harvey ever saw any proof-sheets of his book. Had they come under his eye he would hardly, even in those days of bibliographical insensibility, have passed so many errors in so few pages. It is probable that when the

---

[1] From 'William Harvey, Physician and Biologist', by Dr H. P. Bayon, in *Annals of Science*, vol. III (1938), p. 62.

[2] *William Harvey* (New York, 1929), p. 99.

first copies reached him he was displeased, and sent Fitzer a list of errata as soon as he was able. A minority of the surviving copies of the book contain an additional half-sheet of two leaves, the first of which carries a list of 126 corrections. Probably these leaves were added only to the remainder of the edition, after the greater part had been distributed— even so, the *errata* list was not long enough; the Latin text edited by Mark Akenside for the College of Physicians in 1766 contains 246 emendations.

The first edition of *De Motu Cordis* is in other respects unsatisfactory, for the greater part of the edition was printed on thin paper of very inferior quality, which has long since turned brown and begun to crumble. Most copies of the book are for this reason very unattractive objects, though there are a few printed on a different paper of excellent quality. The finest copy of this kind that I have seen is the one in the Hunterian collection at the Glasgow University Library. It is sad that so few examples of a book of such great historical significance seem destined to survive in a recognizable form.

Later editions were printed in Leyden, Padua and Rotterdam, and it was not until 1660 that an edition was printed in London. This edition is distinguished in that it was the first after the rare Padua edition of 1643 to contain the approving letters of Johannes Walæus, Professor of Anatomy at Leyden.

The engravings with which Harvey illustrated his remarks upon the venous circulation and the valves in veins were not altogether original. The valves in veins had already been studied and described by several observers including Fabricius, as already mentioned. The tract by Fabricius entitled *De Venarum Ostiolis*, published in 1603 and again in 1624, contained the first adequate illustrations of the valves. It is clear that the artist who drew Harvey's figures did so with the plate from Fabricius before him, and he followed it very closely. A man's forearm is represented arranged as for blood-letting. The hand grasps a barber's pole, and the upper arm is bound with a tourniquet so as to distend the veins below. Both engravings are reproduced here for comparison. Harvey did not make any acknowledgment of his debt, but the work of Fabricius was no doubt so well known to his contemporaries that none was needed.

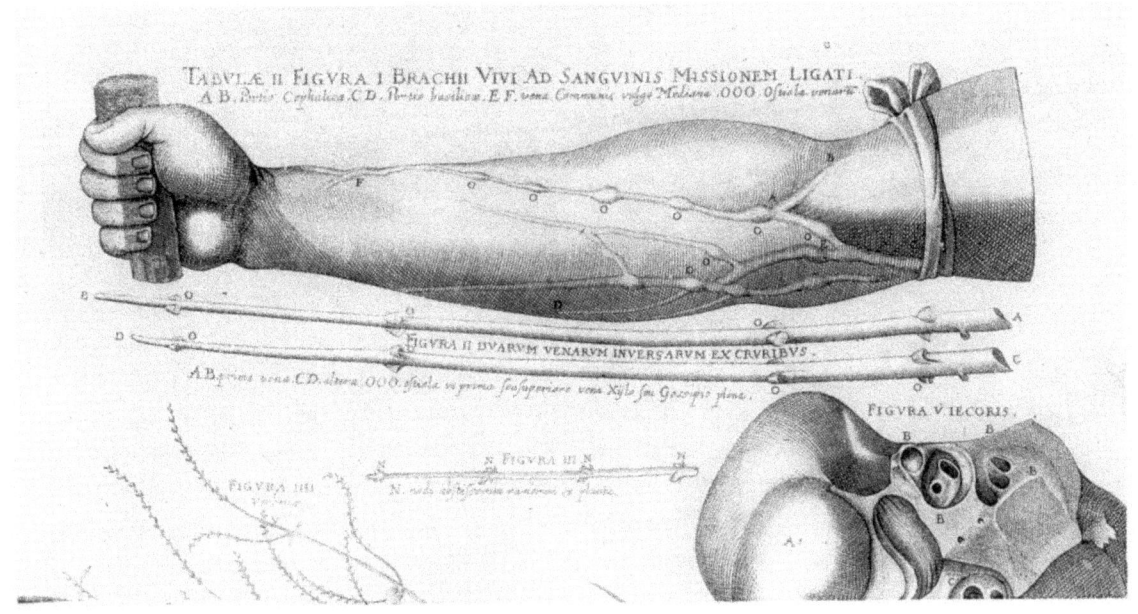

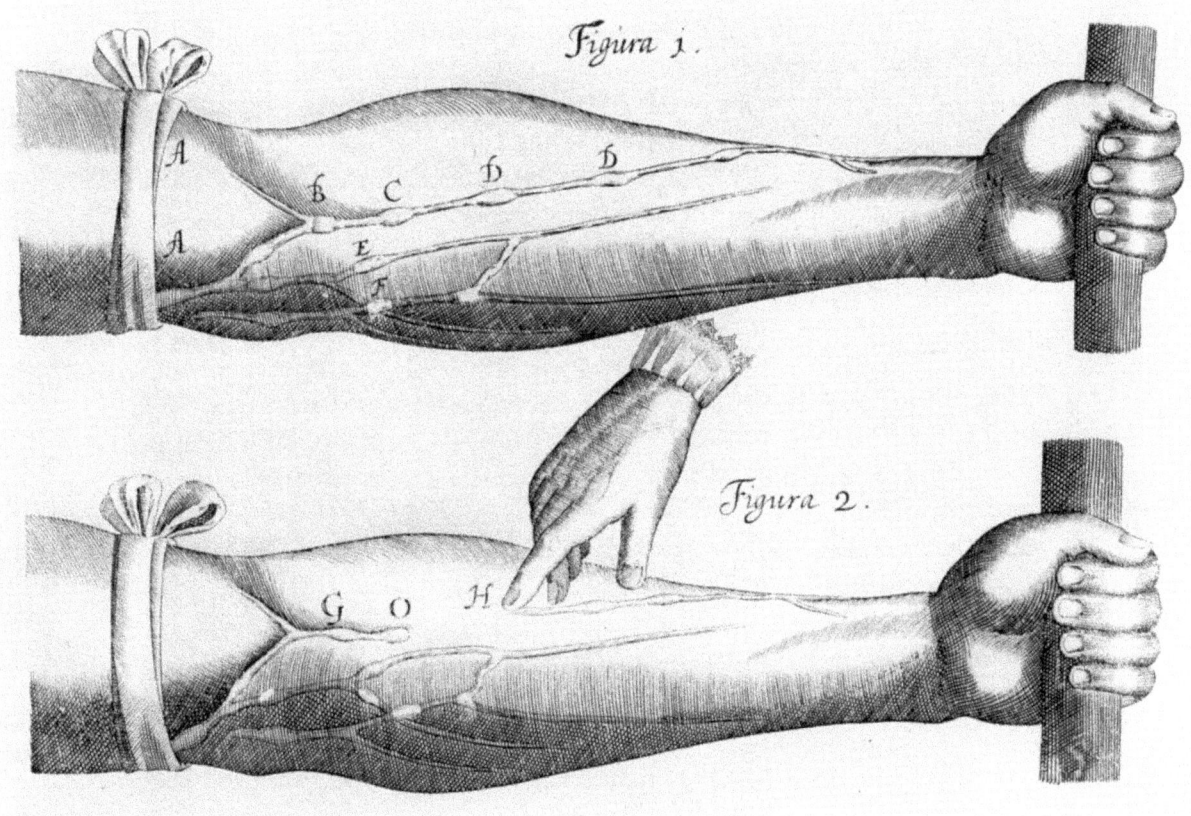

2. VALVES IN VEINS

Above: after Fabricius, about half the size of the original engraving. Below: after Harvey

The first edition of *De Motu Cordis* is not an extremely rare book, forty-six copies being recorded below, but it has become very difficult to obtain. The medical libraries and private collectors have absorbed the more easily available copies, and it is many years since an example was sold in a London auction room.

The Latin text has been printed twenty-three times, including its appearances in the collected works and in facsimiles of the first edition. The book has also been printed eleven times in English, once in Dutch, twice in German, once in Danish, thrice in French, twice in Spanish, and twice in Russian. The Dutch translator included in his edition of 1650 a poem on the death of the author seven years before he died, a grim, if unintentional, jest, which Harvey no doubt knew how to appreciate. The English editions include four different versions. The first of these was made during Harvey's lifetime, though he himself probably had no hand in it. It was published in 1653 with the additions by the Dutch physicians Zachariah Wood and James de Back, and again in 1673, but was not reprinted until 1928, when I used it for the Tercentenary Edition published by the Nonesuch Press. It gives a vigorous, if unpolished, version of Harvey's book in contemporary language, and by no means merited the oblivion into which it had fallen since 1673. A new translation was made by Dr Michael Ryan for his *London Medical and Surgical Journal* (1832–3), and another by Dr Robert Willis for the Sydenham Society's edition of the *Works* published in 1847. The most recent was made for American readers by Professor Chauncey D. Leake, and was published in 1928, 1931, and 1941. Willis's translation had for many years been accepted in England and America as the standard version, though its style is stilted and undistinguished and it is not always more accurate than the earlier text it was designed to supersede.

Most of the Latin editions contained engraved, woodcut, or lithographed copies of the two plates in the first edition. For some reason unknown, the English translations of 1653 and 1673 omitted the figures altogether, although the references in the text were allowed to stand. In the Nonesuch edition of 1928 this defect was remedied, a copper-plate engraving of real beauty being made for it from a drawing by Stephen Gooden, R.A. This is reproduced here as it was the first time since 1628 that the illustrations of Harvey's treatise could be regarded as possessing individual merit.

The majority of the editions of *De Motu Cordis* were recorded by Charles Perry Fisher in his list published in the *Transactions* of the College of Physicians of Philadelphia in 1912 (see no. 53). Several additions were made in the first edition of this *Bibliography*, the most important being those of 1635 and 1689. It was surprising that the earlier of these had not been recorded before, since it was fully described in a bibliographical note inserted in Maire's edition of 1639. It was very unusual for a seventeenth-century publisher to be at the pains to acknowledge his predecessors in this way, and it must have been the more galling to the shade of John Maire that his careful note should have been so consistently ignored. The edition published at Padua in 1689 (no. 12) seems to be extremely rare. In 1928 I recorded only two copies, in the Osler Library and in my collection, and only two others have come to light since then. Another very rare edition is that published at Bologna in 1697 (no. 13). In 1928 only one copy was known, though five more have now been added, together with a full collation from the copy in my collection. In this revised bibliography I am able to record only one edition previously unknown from the collection of Dr Erik Waller, a reissue in 1753 of the Leyden edition of 1736—itself an uncommon book, though it is only incidentally a separate edition, since the sheets of the greater part of the issue were used in the *Opera* of 1737 (see no. 46).

Harvey's treatise has not been printed in Latin with ordinary types since 1824. The scientific world is now unable to read Latin easily, and it seems unlikely that any publisher will risk his money upon a venture so little likely to be appreciated. Photographic facsimiles of the first edition have, however, been made to accompany some of the translations into other languages.

The list of copies given here has been compiled with the kind help of Dr Ernst Weil, who recorded most of them in his article already cited.

EXERCITATIO
ANATOMICA DE
MOTV CORDIS ET SAN-
GVINIS IN ANIMALI-
BVS,

GVILIELMI HARVEI ANGLI,
Medici Regii, & Professoris Anatomiæ in Col-
legio Medicorum Londinensi.

FRANCOFVRTI,
Sumptibus GVILIELMI FITZERI
ANNO M. DC. XXVIII.

PLATE 3

1   DE MOTU CORDIS   4°   1628

Title: Exercitatio Anatomica de Motu Cordis et Sanguinis in Animalibus, Guilielmi Harvei Angli, Medici Regii, & Profefforis Anatomiæ in Collegio Medicorum Londinenfi. [*engraved vignette* 9 × 11 cm.] Francofurti, Sumptibus Guilielmi Fitzeri. [*rule*] Anno M.DC.XXVIII.

Collation: A–I⁴ K²; 38 leaves.

Contents: A1 title; A2*a*–*b* (pp. 3–4) *Dedicatio Carolo Regi*; A3*a*–B1*a* (pp. 5–9) *Dedicatio D.D. Argent*; B1*b*–C2*a* (pp. 10–19) *Procemium*; C2*b*–I4*b* (pp. 20–72) *De Motu Cordis*; K1*a*–*b* errata; K2 blank.

Illustrations: Two engraved plates, illustrating the veins and their valves in the forearm (figures 1–4), inserted between G4 and G5 facing p. 56. Size of plate-mark, 11 × 16·5 cm.

Note: The engraving on the title-page is a device of William Fitzer, the publisher, and carries his monogram (see reproduction). It is of somewhat disproportionate size, and the platemark overlaps the typography above and below. The figures on the plates are based upon one in the *De venarum ostiolis* of Fabricius, 1603 (see reproductions). The half-sheet K is missing from many copies, and was probably added only to part of the edition. Its presence or absence is recorded as far as possible in the following list of copies.

Copies:
(a)   WITHOUT ERRATA LEAF K1
BLO (three copies, one lacking A2), BWU, CCC, LUM, NYAM, RCP (lacking A2), RCS, SGL, ULE (two copies), WHML, YML.
Aberdeen, King's College Library; Bibliothèque National, Paris; Buffalo Museum of Science; Cambridge, Trinity College Library; Chapin Library, Williamstown, Mass.; Columbia University Library; Erlangen University Library; Johns Hopkins Medical School Library; Mazarine Library, Paris (library of Jean de Cordes, d. 1642); Yale University Library.
Dr Logan Clendening, Dr McGraw, Dr L. Reynolds, all U.S.A.

(b)   WITH ERRATA LEAF K1
BM (two copies), BML, CPP (on thick paper), FMP, GUL (on thick paper), OL, RSM.
Dr R. Levy, Dr J. C. Trent (on thick paper), both U.S.A.

(c)   ERRATA LEAF K1 NOT RECORDED
Bonn, Breslau, Göttingen, Jena, Munster University Libraries. Landau copy, now in U.S.A.; Dr Martin Bodmer, Zürich; a private collection in New York.

2   PARISANI EXERCITATIONES [DE MOTU CORDIS]

f° 1635

Title: Aemylii Parifani Romani Philofophi, ac Medici Veneti Nobilium
Exercitationum de Subtilitate Pars Altera de Diaphragmate Singularis
Certaminis Lapis Lydius ad Ioannem Riolanum Iuniorem Anatomicum
Parifienfem Medicum Regium. [*vignette*]
Venetiis, Apud Marcum Antonium Brogiollum. MDCXXXV. Superiorum
permiffu, & Privilegiis.

Sub-title: Aemylii Parifani...Nobilium Exercitationum de Subtilitate
Pars Altera de Cordis et Sanguinis Motu Singularis Certaminis Lapis
Lydius. Ad Guilielmum Harveum Anglum Anatomicum Londinenfem
Medicum Regium.

Collation: 3 vols., f°. [Vol. I, 1623; vol. II, 1635; vol. III, 1638.] Vol. II, Ii6, Kk–
Zz⁶, [*]²; 87 leaves.

Contents: Ii6 sub-title as above; Kk1*a*–Kk2*b* (pp. 385–388, misnumbered 386)
Parifani dedicatio; Kk3*a*–Xx4*a* (pp. 387–521) *De Cordis et Sanguinis Motu Singularis Certaminis Lapis Lydius*; Xx4*b* blank; Yy1*a*–[*]1*a Index*; [*]1*b* blank;
[*]2*a Errata*; [*]2*b* blank.

Note: The greater part of Harvey's treatise is printed in paragraphs alternately with
the refutations of Parisanus. The whole of chapter 1 and parts of the *Procemium* and
chapter 16 are omitted. The figures are not included, though Rr3 is blank as if for
their reception.

This edition, although incomplete, is to be regarded as the second printing
of *De Motu Cordis*; it was not recorded by any other authority before 1928. The
complete work was reprinted in a more accessible form in 1639 (see next entry),
the publisher of which inserted a full bibliographical note recording the previous
printings as follows:

'Parifanus duas partes *in folio* edidit *Exercitationum De microcofmica fubtilitate, Venetiis,*
apud Marcum-Antonium Brogiollum, MDCXXXV, fub nomine *Lapidis Lydii,* ex-
cufas. Harum pofterior, poft paginam 382, Harveii libellum refutatum exhibet; eâ
quidem methodo, ut Harveii textus perpartes vel capita ordinatim recenfeatur,
fingulis-que fua fubiungatur refutatio. Contigit tamen, ut inter alia ex procemio
Harveii nonnullæ particulæ, cum capite primo, ac parte decimi-fexti, fint omiffæ
in recenfione Parifani; quæ heic fuis locis, ex exemplari Francofurtenfi quod dixi,
diligenter funt infertæ; atque omnia typis, auctorum-que titulis & nominibus, per-
fpicue & perpetuò, meliùs & emendatiùs ac in Editione Venetâ factum, diftincta.'

Copy: BM.

3   DE MOTU CORDIS CUM REFUTATIONIBUS 4° 1639

Title: Guilielmi Harveii, Angli, Medici Regii, & in Londinenſi medicorum collegio profeſſoris anatomiæ, De Motu Cordis & ſanguinis in
animalibus, Anatomica Exercitatio. Cum refutationibus Æmylii Pariſani, Romani, philoſophi, ac medici Veneti; et Iacobi Primiroſii, in
Londinenſi collegio doctoris medici. [*device*]
Lugduni Batavorum, Ex officina Ioannis Maire. cɪɔ ɪɔc xxxɪx.

Collation: (?)² [*]² A–Z Aa–Kk⁴ Ll² a–k⁴ l²; 180 leaves.

Contents: (?)1 title; (?)2 *a–b Carolo Regi Dedicatio Harveii*; [*]1*a*–2*b Lectori S.*; A1*a*–
A3*b* (pp. 1–6) *Pariſani Dedicatoria*; A4*a*–Ll2*a* (pp. 7–267) *De Motu Cordis*
(*Harveii tactus, Pariſani contactus* 1–200, *Pariſani nodus* 1–56); Ll2*b* blank; a1 sub-
title to *Iacobi Primiroſii Exercitationes & Animadverſiones*; a2*a*–a3*a* (pp. 3–5) *Carolo
Regi Dedicatio*; a3*b* (p. 6) *D. Argent Salutem*; a4*a–b* (pp. 7–8) *Guilielmo Harveio
ſalutem*; b1*a*–12*b* (pp. 9–84) *Animadverſiones*.

Illustrations: Copies of the four figures in *De Motu Cordis* engraved on two plates, the
plate-marks of which measure about 11 × 18 cm. The two leaves on which they are
printed are inserted in many different positions, but they refer to the text on pp. 185–6
(Aa1).

Note: The make-up of this volume was stated in the first edition of this *Bibliography* to
include Gaspar Aselli's work *De Lactibus*, 1640, but this is now seen to be untrue.
Aselli's treatise is mentioned at the end of the publisher's notice *Ad lectorem* as about
to be published, and is not infrequently bound up with Harvey's work, as are other
works by Le Roy (*Spongia*, 1640) and Primrose (*Antidotum adverſus Spongiam*, 1640,
and *Animadverſiones in Wallaei diſputationem*, 1640), all published by Maire; but
none of these is truly part of the volume now described. The position of the two
unsigned leaves *Ad lectorem* is variable; they may follow the first dedication, as
collated here, or they may be at the end of the volume. They form a separate gathering
of two leaves and were placed by the binder as he pleased.

Harvey's tract is printed passage by passage alternately with refutations by Parisanus.
This composite work had already been printed, though with omissions, in a folio
volume published at Venice in 1635 (see no. 2), which was described by Maire on
the leaves *Ad lectorem*. The passages omitted are here restored. The volume of 1639
was reprinted in 1647 with additions (see no. 6).

Copies: BM, BML, BWU, CPP, FMP, GUL, LUM, NYAM, OL, RCP, RCS,
RFG, SGL (two copies¹), ULC, WHML (two copies), YML.
G. L. Keynes.

¹ One of these copies was recorded in the catalogue of the Surgeon-General's Library as having
the date 1739. This has misled some authorities into recording an imaginary edition of this date.

4 DE MOTU CORDIS   12° 1643

Title: Guilielmi Harveji, Angli, Medici Regii, & Anatomici Londinenſis celeberrimi. De Motu Cordis et Sanguinis in Animalibus Anatomica Exercitatio. Cui Poſtremâ hac editione acceſſerunt Cl. V. Johannis Walæi Profeſsoris Leydenſis Epiſtolæ duæ, quibus Harveji Doctrina roboratur. [*device*]
Patavii, Apud Sebaſtianum Sardum. M.DC.XLIII. Sumptibus Dominici Ricciardi.

Collation: ✠⁶ *A*¹² B–I¹² K⁶; 120 leaves.

Contents: ✠ 1 title; ✠ 2*a*–✠4*b* printer's address to *D. Jo: Baptiſta Soncinus*; ✠ 5*a*–✠ 6*a Index capitum*; ✠ 6*b* blank; A1*a*–B1*a* (pp. 1–25) *Prœmium*; B1*b* blank; B2*a*–F8*b* (pp. 27–136) *De Motu Cordis*; F9 sub-title to *Johannis Walæi De Motu Sanguinis Epiſtolæ Duæ*; F10*a*–K6*a* (pp. 139–227) *Epiſtola Prior, Altera*; K6*b* blank.

Illustration: A folding leaf inserted opposite p. 103 with copies of the usual figures engraved on two plates, the plate-marks of which measure 11·5 × 15 cm.

Note: The letters of Walæus are here added for the first time. They were not reprinted until the London edition of 1660.

Copies: BM, BWU, CPP, GUL, LUM, MSL, NYAM, RCP, SGL, YML.
        G. L. Keynes (untrimmed).

SPIGELII OPERA [DE MOTU CORDIS]   f° 1645

Engraved title: Adriani Spigelii Bruxellenſis…Opera Quæ extant, Omnia.

5   Ex recenſione Ioh: Antonidæ Vander Linden…
        Amſterdami, Apud Iohannem Blaev. CIↃ IↃC XLV.

Collation: 2 vols. f°. Vol. 1. Ff⁴ Gg⁴ Hh⁶; 14 leaves.

Contents: Ff1*a* sub-title to *De Motu Cordis*; Ff1*b* dedication to King Charles; Ff2*a* (p. xxxviiij) dedication to D. Argent; Ff2*b*–Ff3*b* (pp. xl–xlij) *Prœmium*; Ff3*a*–Hh6*b* (pp. xliij–lxiiij) *De Motu Cordis*.

Illustration: on Hh3*a* (p. lvij) copies of the usual figures on a plate measuring 33·5 × 21 cm.

Note: An open copy of this book is laid before Harvey in the contemporary portrait now in the Hunterian collection, University of Glasgow (see my *Portraiture of William Harvey*, 1949, p. 32).

Copies: BM, BML, BWU, CPP, NYAM, RCS, SGL, ULE, WHML, &c.

GUILIELMI HARVEJI,
Angli, Medici Regii, & Anatomici
Londinensis celeberrimi.

De
MOTU CORDIS ET SANGUINIS
in Animalibus
ANATOMICA EXERCITATIO,
Cui
Postremâ hac editione accesserunt
Cl. V.
JOHANNIS VVALÆI
Professoris Leydensis
Epistolæ duæ, quibus Harveji
Doctrina roboratur.

PATAVII,
Apud Sebastianum Sardum. M.DC.XLIII.
Sumptibus Dominici Ricciardi.

6 RECENTIORUM DISCEPTATIONES [DE MOTU CORDIS ETC.] 4° 1647

Title: Recentiorum Difceptationes De Motu Cordis, Sanguinis, et Chyli, in Animalibus. Quorum feries poſt alteram paginam exhibetur. [*device*] Lugduni Batavorum, Ex Officina Ioannis Maire. CIƆ IƆ C XLVII.

Collation: *⁴ A–Z Aa–Kk⁴ Ll² a–l⁴ A–Z Aa–Gg⁴ A–F⁴ G²)?(⁴ [*]⁴ A–O⁴; 392 leaves.

Contents: *1 title; *2*a* Typographus Lectoribus S.D.; *2*b* contents; *3*a* sub-title to *Gulielmi Harveii...De Motu Cordis...Cum refutationibus Æmylii Pariſani...et Iacobi Primiroſii*; *3*b* blank; *4 *Carolo Regi Dedicatio Harveii*; A1*a*–A3*b* (pp. 1–6) *Pariſani Dedicatoria*; A4*a*–Ll2*a* (pp. 7–267) *De Motu Cordis*; Ll2*b* blank; a1*a* sub-title to *Iacobi Primiroſii Exercitationes & Animadverſiones*; a1*b* blank; a2*a*–a3*a* (pp. 3–5) *Carolo Regi Dedicatio*; a3*b* (p. 6) *D. Argent Salutem*; a4*a*–*b* (pp. 7–8) *Guilielmo Harveio Salutem*; b1*a*–14*b* (pp. 9–84) *Animadverſiones*; A1*a*–Gg4*b* (pp. 1–240) *Theſes De Circulatione Naturali* (by Walæus, Primirosius, Henricus Regius, Rogerus Drake); A1*a* sub-title to *Iacobi Primiroſii...Antidotum Adverſus Henrici Regii...Spongiam*; A1*b* *Ad Lectorem*; A2*a*–G2*a* (pp. 3–51) *Antidotum*; G2*b* blank;)?(1*a* *De Lactibus...Gaſparis Aſellii...Diſſertatio*;)?(1*b* blank;)?(2*a*–)?(3*b* *Senatui Mediolanenſi Epiſtola Dedicatoria*;)?(4*a* *Typographus Candido Lectori S.*;)?(4*b* *Lectori*; [*]1*a*–[*]4*a* four anatomical plates with explanations facing each; [*]4*b* blank; A1*a*–N4*b* (pp. 1–104) *De Lactibus*; O1*a*–O4*b* *Index*.

Illustrations: Six engraved plates as in the edition of 1639 (see no. 3). There are also two woodcuts on pp. 43 and 50 of the thesis by Walæus.

Note: This volume is a reprint of the edition of 1639 (see no. 3) with several additional theses. The make-up is very complex, and the correct constitution is difficult to elucidate.

Copies: BM, BLO, BML, BWU, CCC, CPP, RSM, WHML.

7 DE MOTU CORDIS 12° 1648

Title: Guilielmi Harvei Doct: & Profeſſ: Regii Exercitatio Anatomica De motu cordis & ſanguinis. Cum Præfatione Zachariæ Sylvii Medici Roterodamenſis. Acceſſit Diſſertatio de Corde Doct. Jacobi de Back, Urbis Roterodami Medici ordinarii. [*ornament*] Roterodami, Ex Officinâ Arnoldi Leers. CIƆ IƆ C XLVIII.

Collation: *¹² **⁸ A–I¹², [*]² A–I¹² K⁸; 148 leaves.

GUILIELMI HARVEI

Doct: & Profeſſ: Regii

Exercitatio Anatomica

De motu cordis & ſanguinis.

Cum Præfatione

ZACHARIÆ SYLVII

Medici Rôterodamenſis.

ACCESSIT

Diſſertatio de Corde

Doct. JACOBI DE BACK,

Vrbis Roterodami Medici ordinarii.

ROTERODAMI,

Ex Officinâ ARNOLDI LEERS.

CIƆ IƆC XLVIII.

Contents: *1 engraved title; *2*a* title; *2*b* lines *In Guilielmum Harveum Zacharias Sylvius*; *3*a*–**1*b* *Zacharaei Sylvii Præfatio*; **2*a*–**3*a* *Dedicatio Carolo Regi*; **3*b*–**7*a* *Dedicatio D. Argent*; **7*b*–**8*b* *Index Capitum*; A1*a*–B1*b* (pp. 1–26) *Proœmium*; B2*a*–I12*a* (pp. 27–215) *De Motu Cordis*; I12*b* blank; [*]1*a* (p. 216) *mendæ sic emendandæ*; [*]1*b* blank; [*]2*a* sub-title to *Jacobi de Back...Differtatio de Corde...Roterodami, Ex Officinâ Arnoldi Leers.* CIↃ IↃC XLVIII.; [*]2*b* *De Harveio et Baccio, Ad Lectorem Epigramma*; A1*a*–*b* (pp. 1–2) *Dedicatio Guilielmo Harveio*; A2*a*–A12*b* (pp. 3–24) *Ad Lectores Alloquium*; B1*a*–K2*a* (pp. 25–219) *Differtatio de Corde*; K2*b*–K5*b* *Sectionum & Capitum Elenchus*; K6*a* *Errata*; K6*b*–K8*b* blank.

Engraved title: an allegorical design representing Truth led by Time. Inscribed on a cloth above: *Guilielmi Harvæi/Medici Regij/Exercitatio Anatomica/de/Cordis et Sanguinis Motu.* Above the figures is a scroll with the motto: *Veritatem Tempus Manuducit.* Inscribed below: *Roterdami,/Apud Arnoldum Leers A° 1648.* Plate-mark 11 × 6 cm.

Illustrations: On G6*b* and G7*a* (pp. 156–7) reduced copies of the plates inserted in no. 1 containing *Figuræ* 1–4. The plate-marks measure 10 × 6 cm.

Note: The first Rotterdam edition of *De Circulatione Sanguinis*, Leers, 1649, is sometimes bound in at the end.

Copies: BM, BLO, BML, BWU, CCC (two copies), CPP, FMP, LUM, MSL, RCP (two copies), RSM (lacks first two leaves), SGL, ULC (lacks engraved title), ULE, WHML (two copies).
    G. L. Keynes, F. C. Pybus.

8   DE MOTU CORDIS: DE CIRCULATIONE SANGUINIS
12° 1654

Title: Guilielmi Harveji Doct: & Profeff: Regii Exercitationes Anatomicæ, De motu Cordis & Sanguinis Circulatione. Cum duplici Indice Capitum & Rerum. Acceffit Differtatio de Corde Doct. Jacobi de Back, Vrbis Roterodami Medici ordinarii. [*ornament*]
Roterodami, Ex Officinâ Arnoldi Leers. CIↃ IↃ CLIV.

Collation: *12 **4 A–M12 N8, A–L12 M6; 306 leaves.

Contents: *1 engraved title; *2*a* title; *2*b* lines *In Guilielmum Harvejum Zacharias Sylvius*; *3*a*–*11*b* *Zachariæ Sylvii Præfatio*; *12*a*–*b* *Dedicatio Carolo Regi*; **1*a*–**3*b* *Dedicatio D. D. Argent*; **4*a*–*b* *Index Capitum*; A1*a*–A11*a* (pp. 1–21) *Proœmium*; A11*b*–H2*b* (pp. 22–172) *De Motu Cordis*; H3*a*–M11*a* (pp. 173–285) *Exercitationes Duæ De Circulatione Sanguinis*; M11*b*–N8*b* *Index rerum*; errata at bottom of N8*b*. A1*a* sub-title to *Differtatio De Corde*; A1*b* *De Harvejo et*

Baccio Ad Lectorem Epigramma; A2a–b (pp. 3–4) *Dedicatio Guilielmo Harvejo*; A3a–C3b (pp. 5–54) *Ad Lectores Alloquium*; C4a–L6b (pp. 55–252) *Diſſertatio de Corde*; L7a–L9b *Sectionum et Capitum Elenchus*; L10a–M6b *Index Rerum*; errata at bottom of M6b.

Engraved title: The same plate as in no. 7, with the inscription above altered to *Guilielmi Harvæi*/*Medici Regij*/*Exercitat. Anatomicæ*/*de Motu Cordis et*/*Sanguinis Circulo.* The date below is altered to 1654. The plate-mark measures 11 × 6 cm.

Illustrations: The same plates as in no. 7 on F3b and F4a (pp. 126–7).

Note: Page 3 of the *Diſſertatio de Corde* is numbered *Pag: 1.* The *Exercitationes Duæ De Circulatione Sanguinis* were printed with *De Motu Cordis* for the first time in this edition. The left-hand headlines for the three treatises are respectively *Exercitatio Anatomica* I, II, and III.

Copies: BM, BML, BWU, FMP, NYAM, RCP, RCPE, RCS (lacks engraved title and last six leaves), RSM, StBH, SGL, ULE (lacks *Dissertatio de Corde*), YML (lacks *Dissertatio de Corde*).

9 DE MOTU CORDIS: DE CIRCULATIONE SANGUINIS
12° 1660

Title: Guilielmi Harveji...Exercitationes Anatomicæ, De motu Cordis & Sanguinis Circulatione...[etc. as in no. 8]
Roterodami, Ex Officinâ Arnoldi Leers. CIↃ IↃ CLX.

Collation, contents: As in no. 8. The lists of errata are omitted.

Engraved title: The same plate as in no. 8, with the date altered to 1661.

Illustrations: The same plates as in no. 8.

Copies: BM, BML, BWU, CCC, CPP, FMP, LUM, MSL (lacks *Diſſertatio de Corde*), NYAM, RFG, SGL, ULC (lacks engraved title), WHML (two copies), YML. G. L. Keynes.

10 DE MOTU CORDIS: DE CIRCULATIONE SANGUINIS
12° 1660

Title: Guilielmi Harveii Doct. & Profeſſ. Regii Exercitationes Anatomicæ, De motu Cordis & Sanguinis Circulatione. Cum duplici Indice, Capitum & Rerum. Quibus acceſſerunt Jo. Walæi, de Motu Chyli & Sanguinis, Epiſtolæ Duæ. Itemque Diſſertatio de Corde Doct. Jac. de Back, Medici Roterodamenſis. *[ornament]*
Londini, Ex Officina R. Danielis, CIↃ IↃC LX.

Collation: *¹² **⁸ A–V¹² X⁴; 264 leaves.

Contents: *1a title; *1b lines *In Guilielmum Harveium Zacharias Sylvius*; *2a–*8b *Zachariæ Sylvii Præfatio*; *9a *Carolo Regi Dedicatio*; *9b–*10b *D. D. Argent Dedicatio*; *11a–b *Index Capitum*; *12a–**8b *Index Rerum*; A1a–A8b (pp. 1–16) *Procemium*; A9a–E12b (pp. 17–120) *De motu Cordis*; F1a–I4a (pp. 121–199) *Exercitationes Duæ*; I4b blank; I5a sub-title to *Johannis Walæi Epistolæ Duæ*; I5b quotation from *Aristoteles, 3 de Hist. Animal., cap.* 2; I6a–M9b (pp. 203–282) *Epistolæ Duae*; M10a sub-title to *Jacobi de Back Dissertatio*; M10b *De Harveio et Baccio Ad Lectorem Epigramma*; M11a–b (pp. 285–6) *Guilielmo Harveio Dedicatio*; M12a–O6a (pp. 287–323) *Ad Lectores Alloquium*; O6b–V4b (pp. 324–464) *Dissertatio de Corde*; V5a–X4a *Sectionum & Capitum Elenchus*; X4b blank.

Engraved title: Inserted before printed title, a copy of the engraving in no. 7, inscribed below: *Londini/Ex Officina R: Danielis/*1661. The plate-mark measures 11 × 6 cm.

Illustrations: On a folding plate inserted before A1 or in sheet D, copies on one plate of the two engravings on pp. 156–7 of no. 7. References are given to pp. 87 and 90. The plate-mark measures 12·5 × 11 cm.

Note: Pp. 87 and 88 are blank. They are allowed for in the pagination although no text is missing, so that they were evidently intended to receive the illustrations, which were actually inserted on a folding plate. Pp. 134,5 and pp. 142,3 have been transposed.

Copies: BWU, CPP, GUL (two copies), NYAM, RCP, RCS, ULC, WHML (four copies, one lacking engraved title, one lacking plate).
G. L. Keynes, F. C. Pybus.

11 DE MOTU CORDIS: DE CIRCULATIONE SANGUINIS
12° 1671

Title: Guilielmi Harveji...Exercitationes Anatomicæ, De motu Cordis & Sanguinis Circulatione...[etc. as in no. 8]
Roterodami, Ex Officinâ Arnoldi Leers. M. DC. LXXI.

Collation, contents: As in no. 9.

Engraved title: As in no. 9. The date has not been altered.

Illustrations: The same plates as in no. 8.

Note: Instead of the floral ornament which was used on the title-pages and sub-titles in the previous editions (nos. 8 and 9) is a rectangular allegorical design with the motto *Labore et Vigilantia*.

Copies: BM, BWU, CPP, GUL (two copies), MSL (two copies, one lacking *Dissertatio de Corde*), RCP (two copies), RCS, SGL, ULC, WHML (three copies), YML.
G. L. Keynes, F. C. Pybus.

GVILIELMI HARVEII,

Angli, Medici Regii, & Anatomici
Londinensis celeberrimi.

DE MOTV CORDIS ET SANGVINIS
in Animalibus
ANATOMICA EXERCITATIO,
Cui
Postrema hac editione accesserunt
Cl. V.

IOHANNIS VVALÆI

Professoris Leydensis
Epistolæ duæ, quibus Harveii
Doctrina roboratur.

PRÆCLARiSSIMO,

Atque Excellentiss. Viro Comiti,
& Equiti Domino

IOANNI PAVLO

GARDINO

Urbis Saccilli, necnon Almę Romæ Nobili, Philosophię ac Medicinæ
Doctori eximio · Dic.

PATAVII, M DC LXXXIX.

Apud Cadorinum,
Superiorum Permissu.

12   DE MOTU CORDIS   12°   1689

Title: Guilielmi Harveii, Angli, Medici Regii, & Anatomici Londinenſis celeberrimi. De Motu Cordis et Sanguinis in Animalibus Anatomica Exercitatio, Cui Poſtrema hac editione aceſſerunt Cl. V. Iohannis Walæi Profeſſoris Leydenſis Epiſtolæ duæ, quibus Harveii Doctrina roboratur. [*rule*] Illuſtriſſimo Viro Comiti, & Equiti Domino Ioanni Paulo Gardino Urbis Saccilli, necnon Almę Romæ Nobili, Philoſophię ac Medicinæ Doctori eximio Dic. [*ornaments*] Patavii, ᴍᴅ ᴄʟxxxɪx. [*rule*] Apud Cadorinum, Superiorum Permiſſu.

Collation: ✠¹² A–G¹² H⁶; 104 leaves.

Contents: ✠1 title; ✠2*a*–✠3*b* dedication by *Iacobus Cadorinus* dated *Batauii* 4. *Iulii* 1689; ✠4*a*–✠12*b Procœmium*; A1*a*–E7*b* (pp. 1–110) *De Motu Cordis*; E8 sub-title to *Cl. V. Iohannis Walæi . . . Epiſtolæ Duæ Ad Thomam Bartholinum . . .*, E9*a*–H4*a* (pp. 113–175) *Epiſtola prior, altera*; H4*b*–H5*b Index Capitum*; H6 blank.

Note: An alteration was made in the title-page while the book was in the press, so that in two of the four copies recorded the words below the first rule appear as: *Præclariſſimo, Atque Excellentiſ. Viro Comiti*, instead of *Illuſtriſſimo Viro Comiti*. This edition, which is rare, does not contain the illustrations.

Copies: BWU (second form of title-page), OL, RCP, YML.
      G. L. Keynes (second form of title-page).

13   DE MOTU CORDIS   12°   1697

Title: Guilielmi Harveii Angli, Medici Regij, & Anatomici Londinenſis celeberrimi. De motu Cordis, & Sanguinis in Animalibus, Anatomica Exercitatio, Cui Poſtrema hac editione acceſſerunt Clariſſimi Viri Iohannis Walæi Profeſſoris Leydenſis Epiſtolæ duæ, quibus Harveii Doctrina roboratur. Bononiæ, 1697. [*rule*] Typis Longhi. Superiorum permiſſu.

Collation: ✠¹² A–G¹² H⁶; 102 leaves.

Contents: ✠1 blank; ✠2 half-title; ✠3 title; ✠4*a*–✠12*b Procœmium*; A1*a*–E7*b* (pp. 1–110) *De Motu Cordis*; E8 sub-title to *Johannis Walæi Epistolæ Duæ ad Thomam Bartholinum*; E9*a*–H4*a* (pp. 113–175) *Epistola prior, altera*; H4*b*–H5*b* (pp. 176–178) *Index Capitum, Imprimatur*; H6*a*–*b* figures.

GULIELMI HARVEI

Angli, Anatomiae ac Chirurgiae in Collegio Medic. Lond. Professoris;
Serenissimaeque Majestatis Regiae Archiatri

EXERCITATIO ANATOMICA
DE MOTU CORDIS

ET SANGUINIS IN ANIMALIBUS.

Cui accedunt ejusdem Auctoris

EXERCITATIONES DUAE ANATOMICAE

DE

CIRCULATIONE SANGUINIS

Ad JOANNEM RIOLANUM Filium;

In Academia Parisiensi Anatomes & Herbariae Professorem Regium,
Reginae Matris Ludovici XIII. Medicum Primarium.

Atque hisce

PRAEFATIONEM

addidit

BERNARDUS SIEGFRIED ALBINUS,

Anatomes & Chirurgiae, in Acad. Lugd. Batav. Profess.

LUGDUNI BATAVORUM,
Apud JOHANNEM van KERCKHEM, 1736.

Illustrations: Crude woodcut copies of the usual figures printed on both sides of the last leaf (H6), with reference to p. 77.

Note: This edition is exceedingly uncommon, though five more copies have come to light since 1928, when only the copy at the College of Physicians, Philadelphia, was known.

Copies: BWU, CPP, NYAM, WHML, LCW (lacking figures).
G. L. Keynes.

14 DE MOTU CORDIS: DE CIRCULATIONE SANGUINIS
$$4° \quad 1736$$

Title: Gulielmi Harvei Angli, Anatomiae ac Chirurgiae in Collegio Medic. Lond. Profefforis; Sereniffimaeque Majeftatis Regiae Archiatri Exercitatio Anatomica De Motu Cordis et Sanguinis in Animalibus. Cui accedunt ejusdem Auctoris Exercitationes Duae Anatomicae De Circulatione Sanguinis Ad Joannem Riolanum Filium; In Academia Parifienfi Anatomes & Herbariae Profefforem Regium, Reginae Matris Ludovici XIII. Medicum Primarium. Atque hifce Praefationem addidit Bernardus Siegfried Albinus, Anatomes & Chirurgiae, in Acad. Lugd. Batav. Profeff. [*engraved vignette*]
Lugduni Batavorum, Apud Johannem van Kerckhem, 1736.

Collation: $*^4 **^4$ A–X^4 Y^2; 94 leaves.

Contents: *1 title; *2a–**3a *Bernardi Siegfried Albini Præfatio ad medicinæ studiosos*; **3b–4b blank; A1a–N4a (pp. 1–103) *De Motu Cordis* with the dedications; N4b blank; O1 sub-title to *Exercitationes Duæ de Circulatione Sanguinis*; O2a–X4a (pp. 107–167) *Exercitationes Duæ*; Y4b blank; Y1a–b (pp. 169–170) *Libri quos excudit...J. A. Kerckhem*; Y2 blank.

Note: This form of van Kerckhem's edition is rare, the sheets of the greater part of the issue having been published in 1737 as Part 1 of the *Opera* (no. 46).

Copies: CPP, MGU, SGL, YML.

14a DE MOTU CORDIS: DE CIRCULATIONE SANGUINIS
$$4° \quad 1753$$

Title: Gulielmi Harvei...Exercitatio Anatomica De Motu Cordis... Exercitationes Duae Anatomicae De Circulatione Sanguinis...[etc. as in no. 14] [*ornament*] Lugduni Batavorum, Apud Gerardum Potvliet, et Cornelium de Pecker. 1753.

GULIELMI HARVEI

Angli, Anatomiae ac Chirurgine in Collegio Medic. Lond. Professoris;
Serenissimaeque Majestatis Regiae Archiatri

EXERCITATIO ANATOMICA

DE MOTU CORDIS

ET SANGUINIS IN ANIMALIBUS.

Cui accedunt ejusdem Auctoris

EXERCITATIONES DUAE ANATOMICAE

DE

CIRCULATIONE SANGUINIS

Ad JOANNEM RIOLANUM Filium;

In Academia Parisiensi Anatomes & Herbariae Professorem Regium,
Reginae Matris Ludovici XIII. Medicum Primarium.

Atque hisce

PRAEFATIONEM

addidit

BERNARDUS SIEGFRIED ALBINUS,

Anatomes & Chirurgiae, in Acad. Lugd. Batav. Profess.

LUGDUNI BATAVORUM,

Apud {GERARDUM POTVLIET,} 1758.
ET
{CORNELIUM DE PECKER.}

Collation, contents: As in no. 14.

Note: This is a reissue of the sheets of no. 14 with a cancel title. Its existence was reported to me by Dr Erik Waller in 1946, but no other copy has come to my notice since then.

Copy: BWU.

15 DE MOTU CORDIS ET SANGUINIS CIRCULATIONE
12° 1751

Title: Guilielmi Harveii Doct. et Profeff. Regii Exercitationes Anatomicae, De Motu Cordis et Sanguinis Circulatione.
Glasguae: In Aedibus R. Urie, Sumptibus D. Baxter, Bibliopolae. MDCCLI.

Collation: a⁶ A–M¹² N⁶; 156 leaves.

Contents: a1 title; a2a–b (pp. iii–iv) *Dedicatio Carolo Regii*; a3a–a5b (pp. v–x) *Dedicatio D. D. Argent*; a6a–b *Index Capitum*; A1a–A12a (pp. 1–23) *Prooemium*; A12b blank; B1a–G12b (pp. 25–168) *De Motu Cordis*; H1a–M7b (pp. 169–278) *Exercitationes duae*; M8a–N6a (pp. 279–299) *Index Rerum*; N6b blank.

Illustrations: Copies of the usual figures on a folding plate inserted before p. 1. The plate-mark measures about 12·5 × 17·5 cm.

Copies: BM, BLO, BWU, CPP, GUL, LUM, MSL, NYAM, OL, RCP, RCPE, RFG, RSM, SGL, WHML (two copies), YML.
G. L. Keynes, F. C. Pybus.

16 DE MOTU CORDIS ET SANGUINIS   8° 1824

Title: Guilielmi Harveii Exercitationes De Motu Cordis et Sanguinis; quas notis pauculis instruendas curavit Thomas Hingston, M.D., Societatis Regiae Medicae Edinburgensis Socius, nunc ex Collegio Reginae Cantabrigiensi. [*rule*]
Edinburgi: Ven. apud Joannem Carfrae et Filium; atque Longman et Socios, Londini. [*rule*] M DCCC XXIV.

Collation: a² b¹ A–P⁸ Q⁴ R¹; 134 leaves.

Contents: a1 half-title; a2 title, imprint *Edinburgi: Excudit Joan. Brewster* on verso; a3a dedication to Andrew Duncan; a3b blank; a4a–a8a (pp. vii–xv) *Lectori S* (editor's introduction); a8b instructions for placing the plates; b1a–b *Argumenta Capitum*; A1 sub-title; A2a–K7a (pp. 3–157) *De Motu Cordis*; K7b blank; K8 sub-title to *Exercitationes Duae*; L1a–R1b (pp. 161–250) *Exercitationes Duae*.

Illustrations: Two plates inserted between H3 and H4 (pp. 118–19). The usual figures redrawn and engraved by Lizars. Plate-marks about 22 × 14 cm.

Note: Printed on ribbed paper and bound in quarter green cloth with paper label on spine; it is also found in brown paper boards. The leaves signed b and R are probably the two halves of a quarter sheet.

Copies: BM, BML, BWU, CPP, NLS, NYAM, OL, RCP, RSM, SGL, ULE, WHML (two copies).
G. L. Keynes.

17 DE MOTU CORDIS 4° 1928

Title: Exercitatio Anatomica De Motu Cordis. . .[etc. as in no. 1].

Colophon: This edition of William Harvey, Exercitatio Anatomica De Motu Cordis et Sanguinis in Animalibus, Francofurti, 1628, was published in 1928, the third centenary of its first publication, by R. Lier & Co., Florence, Lungarno Torrigiani 19. 250 copies of this publication were printed for the Royal College of Physicians of London.

Collation: 21·2 cm., pp. [ii] + 72 + [2].

Illustrations: Two plates as in no. 1 inserted between pp. 56, 57.

Note: The copy from which this collotype facsimile was made seems to have lacked the half-sheet K, so that the facsimile is without the *errata*. Bound in boards with paper labels on spine and side.

17a DE MOTU CORDIS 4° 1928

Title: Exercitatio Anatomica De Motu Cordis. . .[etc. as in no. 1].

Colophon: This facsimile edition of Exercitatio Anatomica De Motu Cordis 1628, by William Harvey, was published in 1928, three hundred years after the first edition of this standard-work of modern medicine, by R. Lier & Co., Florence, 19 Lungarno Torrigiani. Being Vol. v of Monumenta Medica edited by Henry E. Sigerist.

Collation: 21·2 cm., pp. 72 + [1].

Illustrations: Two plates as in no. 1 inserted between pp. 56, 57.

TRANSLATIONS

18 DE MOTU CORDIS: *DUTCH*   8°   1650

Title: Gulielmus Harvejus, Hoog leer-meefter, ende genees-heer des
    Konings van Engelant, Vande Beweging van't Hert, Ende Bloet. Vit
    het Latijn vertaalt door N. van Affendelft, Ende nu Tot nut en voordeel
    van alle Chirurgijns, en Lief-hebbers in't licht gebracht. [*ornament*]
    t'Amfteldam, [*rule*] Voor Cornelis Laft, Boek-verkoper inde Gaft-huis-
    meule fteeg, inde vier Euangeliften. cıɔ ıɔ cl.

Collation: *⁸ A–G⁸ A–D⁸ Aa⁸ Bb⁴; 108 leaves.

Contents: *1 engraved title; *2 printed title; *3a–b verses by *N.V.A.[ffendelft]* headed
    Op de Doot; *4a–A5b *Voor-rede*; A6a–G6a (pp. 1–97) *Beweging van't Hert*;
    G6b–G7b *Blatwyzer*; G8 blank. A1 sub-title to: *Twee Brieven Van de Beweginge
    des Chyls Ende des Bloeds. Befchreven door...D. Joh. Walaeus...t' Amstelredam.
    Voor Cornelis Laft...1650*; A2a–Bb4b (pp. 1–62, 1–24) *Den Eerften Brief, De
    tweede Brief.* At bottom of Bb4b: *t'Amfteldam, Gedrukt by Adriaan Roeft, Boek-
    drukker, woonende voor aan inde Bloemftraat, 1650.*

Engraved title: A copy of the plate in the Latin edition of 1648 (no. 7), inscribed
    above: *d'Anatomifch Oeffening Van Gulielmus Harveius, Over de Beweging van't
    Hert ende Bloed.* The scroll over the figure bears the words: *Den tyt is de Baar-
    moeder aller Dinggen.* Inscribed below: *'t Amfteldam. Voor Cornelis Laft, A°.1650.*

Note: In the copy collated the printer has transposed the type on G6b and G7a. The
    verses on the third leaf are headed 'On the Death' of Harvey. There is no reason
    for supposing the date 1650 on the title-page and elsewhere to be erroneous, and we
    must therefore suppose that the translator had been misinformed.

Copies: BWU, CPP, OL, SGL.

19 DE MOTU CORDIS: DE CIRCULATIONE SANGUINIS:
    ENGLISH   8°   1653

Title: The Anatomical Exercifes of Dr William Harvey Profeffor of
    Phyfick, and Phyfician to the Kings Majefty, Concerning the motion of
    the Heart and Blood. With the Preface of Zachariah Wood Phyfician
    of Roterdam. To which is added Dr. James De Back his difcourfe of the
    Heart, Phyfician in ordinary to the Town of Roterdam. [*rule*]
    London, Printed by Francis Leach, for Richard Lowndes at the White
    Lion in St. Pauls Churchyard, near the West end, 1653.

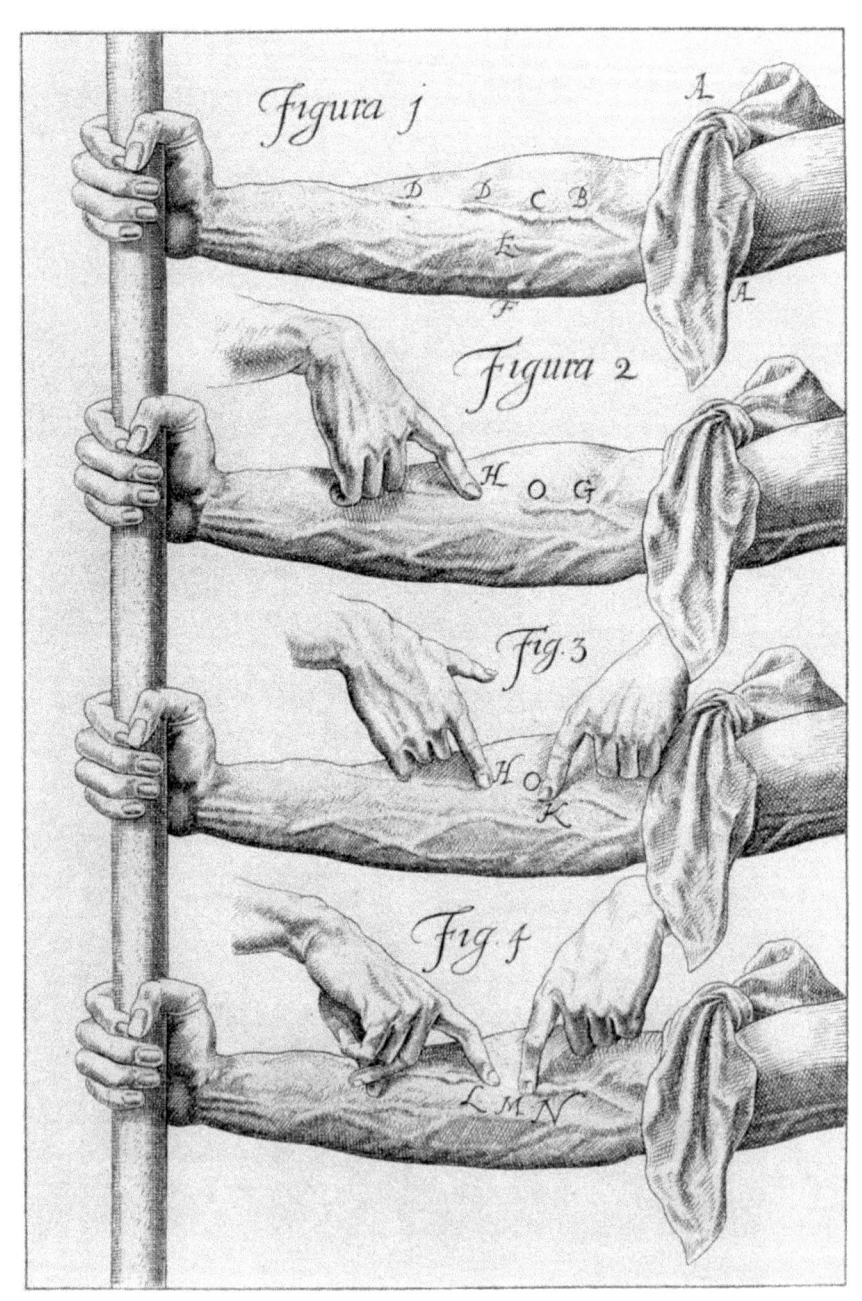

4. *VALVES IN VEINS*

Engraved after a drawing by S. Gooden

GULIELMUS HARVEJUS,

Hoog leer-meester, ende genees-heer
des Konings van Engelant,

Vande

BEWEGING van 't HERT,

Ende

BLOET.

Vit het Latijn vertaalt door N. van Assendelst,

Ende nu

Tot nut en voordeel van alle Chirurgijns,
en Lief-hebbers in 't licht gebracht.

t'AMSTELDAM,

Voor CORNELIS LAST
Boek-verkoper inde Gast-huis-meule
steeg, inde vier Euangelisten.
CIƆ IƆ CL.

Collation: ₓ⁸ *ₓ⁸ *ₓ*⁴ A–X⁸ Y⁴; 192 leaves.

Contents: *1 blank; *2 (cancel leaf) title; *3a–**1a *The Preface of Zachariah Wood*; **1b–2a dedication to King Charles; **2b–4a dedication to Dr Argent; **4b–5a *Index of the chapters*; **5b–***4a *The Proeme*; ***4b blank; A1a–G8a (pp. 1–111) *Anatomical Exercifes*; G8b blank; H1 sub-title to *The Difcourfes of James de Back . . . London, Printed by Francis Leach, 1653*; H2a–I1b *A Speech to the Readers*; I2a–b *To the Reverend and moft Learned Man, William Harvey*; I3a–P8b (pp. 1–108) *Iames de Back, his Difcourfe*; Q1a–Q8a (pp. 109–123) *An Addition [In defence of Harvey's Circulation]*; Q8b blank; R1 sub-title to *Two Anatomical Exercitations Concerning The Circulation of the Blood, To John Riolan the Son, . . . The Author, William Harvey, . . . London, Printed by Francis Leach, 1653*; R2a–Y4b (pp. 1–86) *Anatomical Exercitations to John Riolan.*

Note: In the three paginated sections the following errors occur: (1) 8, 59 for 5, 92, and pagination of p. 100 omitted; (2) 46, 84 for 45, 83; (3) 5, 32 for 3, 23.

In most copies the title-page is a cancel leaf, though sometimes the original title-page has been left in position. In those that I have seen the lower corner is clipped off, so that the date has been removed (see reproduction). This title-page is identical with the other, and is from the same setting of type, except that the rule is omitted above the imprint which reads: *London, Printed by Francis Leach* [? 1653]. On the verso, between two rows of ornaments, are the lines:

ZACHARIE WOOD upon Doctor
WILLIAM HARVEY

Long *Ariftotle*, long may *Galen* live,
Whofe great renown all ages shall furvive,
And long live *Harvey*, they the Arts did find,
Which this brave Englishman now has refind.

Copies: BM, BML, BWU, CPP, NYAM, OL, RCP (two copies, one with two title-pages), RCS, RFG, RSM, SGL, WHML (three copies, two with the two title-pages), YML.
G. L. Keynes, F. C. Pybus.

20 DE MOTU CORDIS: DE CIRCULATIONE SANGUINIS: *ENGLISH* 8° 1673

Title: The Anatomical Exercifes of Dr William Harvey, Profeffor of Phyfick, and Phyfician to King Charles the Firft; Concerning the motion of the Heart and Blood. With The Preface of Zachariah Wood, Phyfician of Roterdam. To which is added, Dr. James de Back, his

THE
ANATOMICAL
Exercifes of

Dr. *WILLIAM HARVEY*

Profeffor of Phyfick,

AND

Phyfician to the Kings Majefty,
Concerning the motion of the *Heart*
and *Blood*.

WITH

The Preface of *Zachariah Wood*
Phyfician of *Roterdam*.

To which is added
Dr. James *De Back* his difcourfe of the
Heart, Phyfician in ordinary to the
Town of *Roterdam*.

LONDON,
Printed by *Francis Leach*

Diſcourſe of the Heart, Phyſician in Ordinary to the Town of Roterdam.
[*rule*]
London, Printed for Richard Lowndes at the White Lion in Duck Lane,
and Math. Gilliflower, at the Sun in Weſtminſter-Hall, 1673.

Collation: a⁴ A–U⁸; 164 leaves.

Contents: a1 title; a2*a–b* dedication to King Charles; a3*a*–a4*b* dedication to Dr Argent;
A1*a*–A7*a The Preface of Zachary Wood*; A7*b* blank; A8*a–b The Index of the Chapters*;
B1*a*–B7*a* (pp. 1–13) *The Proeme*; B7*b* blank; B8*a*–H6*a* (pp. 15–107) *Anatomical
Exerciſes*; H6*b* blank; H7 sub-title to *The Diſcourſe of James de Back...London,
Printed by T. R. 1673*; H8*a*–I7*a A Speech to the Readers*; I7*b*–I8*b To the Reverend
and moſt Learned Man, William Harvey*; K1*a*–P8*a* (pp. 1–91) *James De Back, his
Diſcourſe*; P8*b*–Q6*b* (pp. 92–104) *An Addition* [*In Defence of Harvey's Circulation*];
Q7 sub-title to *Two Anatomical Exercitations Concerning the Circulation of the Blood.
To John Riolan the Son,...The Author, William Harvey,...London, Printed
1673.*; Q8*a*–U8*b* (pp. 107–172) *Anatomical Exercitations to John Riolan.*

Note: The signatures on U3, U4 are misprinted T3, T4. This is a reprint of the
edition of 1653 with some textual errors and omissions.

Copies: BM, BLO, BML, BWU, CPP, LUM, MSL, NYAM, RCP, RCPE,
RCS, RFG, RSM, SGL, ULE, WHML (five copies, one imperfect), YML.

G. L. Keynes, F. C. Pybus.

21   DE MOTU CORDIS: *ENGLISH*   8°   1832

Title: The London Medical and Surgical Journal...Edited by Michael
Ryan, M.D....Vol. I. [II.]
London: Published by Renshaw and Rush, 356, Strand...1832.
[1833.]

Collation: 21·5 cm., vol. I, pp. [ii]+846, vol. II, pp. [ii]+838.

Sub-title: The Anatomical Exercitations of William Harvey, M.D....on the Motion
of the Heart and Circulation of the Blood. Translated from the original Latin, by
M. Ryan, M.D. To which is prefixed, a Biographical Sketch of the Illustrious
Author. (Vol. I, pp. 523–26 Biographical sketch; pp. 556–9, 591–4, 649–52,
685–7, 810–12. Vol. II, pp. 42–5, 197–9.)

Copies: BM, BML, RCS, RSM, WHML.

22 DE MOTU CORDIS: *ENGLISH* 8° 1889

Title: On the Motion of the Heart and Blood in Animals. By William Harvey, M.D. Willis's translation, revised and edited by Alex. Bowie, M.D., C.M.,...[Bohn's Select Library].
London: George Bell and Sons, York Street, Covent Garden. 1889.

Collation: 18 cm., pp. [ii] + xx + 147 + [1], in eights.

Note: Reprinted from the *Works*, 1847, with alterations by the editor.

Copies: BM, BML, BWU, CPP, RSM, SGL, WHML, &c.

23 DE MOTU CORDIS: *ENGLISH & LATIN* 4° 1894

Title: An Anatomical Dissertation upon the Movement of the Heart and Blood in Animals, being A Statement of the Discovery of The Circulation of the Blood. By William Harvey, M.D.,...Privately reproduced in facsimile from the Original Edition printed at Franckfort-on-the-Maine in the year 1628, with a Translation and Memoir, for G. Moreton, 42, Burgate Street, Canterbury. 1894.

Collation: 21·5 cm., pp. x + 72 + [4] + 91 + [1], in fours.

Illustrations: i. Frontispiece. Portrait of Harvey reproduced in photogravure from the painting in the National Portrait Gallery. ii, iii. Facing pp. 57, 58 of the *De Motu Cordis*, facsimile reproductions of the two plates in the first edition of 1628.

Note: The whole of the *De Motu Cordis* of 1628 is reproduced in facsimile, apparently by photo-lithography. The Memoir is signed B., probably the publisher. The English text is altered from that of Willis, though no acknowledgment is made. Bound in quarter parchment and cloth. A hundred numbered copies were printed on large paper.

Copies: BM, BLO, BML, BWU, CPP, NYAM, RCS, SGL, ULC, WHML, &c.

24 DE MOTU CORDIS, ETC.: *ENGLISH* 8° 1907

Title: An Anatomical Disquisition on The Motion of the Heart & Blood in Animals By William Harvey Translated from the Latin by Robert Willis.
London & Toronto: J. M. Dent & Sons Ltd. New York: E. P. Dutton & Co. [1907.]

Collation: 17 cm., pp. [ii] + xxx + 239 + [1], in eights.

Contents: Introduction by E. A. Parkyn.

> An Anatomical Disquisition on the Motion of the Heart and Blood in Animals.
>
> The First Anatomical Disquisition...to John Riolan.
>
> The Second Disquisition to John Riolan.
>
> Letters.
>
> The Anatomical Examination of the Body of Thomas Parr.
>
> The Last Will and Testament of William Harvey, M.D.

Note: Everyman's Library, no. 262, called on the half-title *The Circulation of the Blood.* First printed in 1907 and reprinted in 1923.

24*a*   DE MOTU CORDIS: *ENGLISH*   1909

In: Epoch-making contributions to medicine, surgery and the allied sciences...Collected by C. N. B. Camac, A.B., M.D. With portraits. Philadelphia and London: W. B. Saunders and Company. 1909.

Pages 23–113: An anatomical disquisition on the motion of the heart and blood in animals. By William Harvey, M.D. Willis's translation, revised and edited by Alex. Bowie, M.D., C.M.

25   DE MOTU CORDIS: DE CIRCULATIONE SANGUINIS: *ENGLISH*   8°   1928

Title: The Anatomical Exercises of Dr William Harvey De Motu Cordis 1628: De Circulatione Sanguinis 1649: ¶ The first English text of 1653 now newly edited by Geoffrey Keynes [*device between ornamental bands*] Issued on the occasion of the tercentenary celebration of the first publication of the text of De Motu Cordis

The Nonesuch Press London [1928]

Collation: 20 cm., pp. xvi + 202 + [6], in eights.

Contents: De Motu Cordis.

> De Circulatione Sanguinis.
>
> Additions by the Editor (Bibliographical postscript, Textual notes).

Illustration: Facing p. 87, the usual figures engraved on copper by C. Sigrist after a drawing by Stephen Gooden, R.A. The plate-mark measures 18 × 12 cm. (See reproduction.)

Note: 1500 copies were printed in Holland at the Enschedé Press with the types of Joan Michael Fleischman and Christopher van Dijck. Bound in niger morocco.

The Anatomical
Exercises of Dr. WILLIAM

HARVEY

DE MOTU CORDIS 1628:
DE CIRCULATIONE SANGUINIS 1649:
⟨ *The first English text of* 1653
now newly edited by
Geoffrey Keynes

Issued on the occasion of
the tercentenary celebration of the first
publication of the text of DE MOTU CORDIS
THE NONESUCH PRESS
LONDON

25*a* DE MOTU CORDIS: *LATIN & ENGLISH*   8°   1928

Title: Tercentennial Edition Exercitatio Anatomica De Motu Cordis et
Sanguinis in Animalibus By William Harvey, M.D. With an English
Translation and Annotations by Chauncey D. Leake Professor of
Pharmacology, University of California
Charles C. Thomas Springfield, Illinois   Baltimore, Maryland
M.CM.XXVIII

Collation: 23·5 cm., pp. [xii] + 72 + [2] + 154 + [4], illustrated.

Note: The translation is preceded by a facsimile of the first edition, 1628, which
includes the extra leaf of errata.
   Also issued in Great Britain with a title-page carrying the imprint, *London:
Ballière, Tindall & Cox.* 1928.

Copies: StBH, &c.

25*b* DE MOTU CORDIS: *ENGLISH*   1931

Title: Exercitatio Anatomica De Motu Cordis et Sanguinis in Animalibus
By William Harvey, M.D. An English Translation with Annotations
by Chauncey D. Leake...Charles C. Thomas. Springfield, Illinois
Baltimore Maryland
M.CM.XXXI

24 cm., pp. xii + [iv] + 150 + [4], 9 illustrations.

Note: This is the second edition of Professor Leake's translation taken from the tercentenary volume of 1928 (no. 25*a*) and presented in an inexpensive form for the
benefit of medical students. The book was re-issued in 1941.

Copies: LUM, OL, RCS.

26 DE MOTU CORDIS: *GERMAN*   8°   1878

Wiliam Harvey, der Entdecker des Blutkreislaufs...Culturhistorisch-
medicinische Abhandlung zur Feier des Dreihundertjährigen Gedenk-
tags der Geburt Harveys (1 April 1578) von Dr. Joh. Hermann Baas.
Mit Harveys Bildniss, Facsimile und den Abbildungen des Originals
in Lithographie.
Stuttgart. Verlag von Ferdinand Enke. 1878.

Collation: 25 cm., pp. [xii] + 116.

Note: On pp. 38–112 is a translation into German of *De Motu Cordis*. Inserted at the end is a folding plate with a lithographic copy of the figures in the first edition of 1628.

Copies: BM, &c.

27   DE MOTU CORDIS: *GERMAN*   8°   1910

Title: Klassiker der Medizin herausgegeben von Karl Sudhoff.
William Harvey Die Bewegung des Herzens und des Blutes 1628 Übersetzt und erläutert von Prof. R. Ritter von Töply in Wien Mit vier Abbildungen im Text.
Leipzig Verlag von Johann Ambrosius Barth 1910

Collation: 19 cm., pp. 120, with reproductions of the usual figures on pp. 76, 77.

Note: With an appendix giving extracts: Aus der Anatomie des Realdo Colombo von Cremona.

Copies: BML, BWU, OL, WHML, YML.

27*a*   DE MOTU CORDIS: *GERMAN*   8°   1936

In: Vom Wirken berühmter Ärzte aus vier Jahrhunderten—Theophrast von Hohenheim (Paracelsus), William Harvey, Leopold Auenbrugger, Carl Gustav Carus.
Ludwigshafen, Knoll. 1936

Sub-title: Die Bewegung des Herzens und des Blutes, 1628, Übersetzet und erläutert von Prof. R. Ritter von Töply in Wien. Leipzig Verlag von Johann Ambrosius Barth, 1910.

Note: This translation, occupying pp. [96]–123, is apparently an abridgement of no. 27.

Copy: NYAM.

28   DE MOTU CORDIS: DE CIRCULATIONE SANGUINIS: *FRENCH*   8°   1879

Title: Harvey La circulation du sang Des movements du cœur chez l'homme et chez les animaux Deux réponses à Riolan Traduction Française avec une introduction historique et des notes par Charles Richet. . . .
Paris G. Masson, Éditeur. . . Boulevard St-Germain, 120, ...
M DCCC LXXIX.

Collation: 23 cm., pp. [iv]+iii+[i]+287+[1], in eights.

Copies: BWU, CPP, NYAM, RCP, RCS, RSM, SGL, WHML.

29　DE MOTU CORDIS: *FRENCH*　　　16°　1892

Title: La circulation du sang Des movements du cœur chez l'homme et chez les animaux....

Paris G. Masson,...1892.

Collation: 16°, pp. 128, with two plates.

Copies: BML, BWU, RCP.

29*a*　DE MOTU CORDIS: *LATIN AND FRENCH*　　　8°　1950

Title: Guillaume Harvey Étude Anatomique du Mouvement du Cœur et du Sang chez les animaux Aperçu historique & traduction française par Charles Laubry...G. Doin & Cie 8, Place de l'Odéon Paris 1950

21 cm., pp. 224+[2].

Note: Pp. 57–128 give a facsimile reproduction of the first edition of *De Motu Cordis*, 1620, without the additional leaves, K 1–2, followed by a translation into French by Dr Laubry, pp. 131–223.

29*b*　DE MOTU CORDIS: *RUSSIAN*　　　1927

Note: The first translation of Harvey's treatise into Russian was published in 1927, but no copy seems to have reached this country. The translation was made by K. M. Bykov, Member of the Academy of Sciences, who appended some notes with a preface by himself and another preface by Ivan Pavlov, dated 1924. The book was reprinted with additions in 1948 (see no. 29 *c*).

29*c*　DE MOTU CORDIS: DE CIRCULATIONE SANGUINIS: *RUSSIAN*　　　4°　1948

Title: Вильям Гарвей Анатомическое исследование о движении сердца и крови у животных [*ornament*] Перевод, редакция и комментарии академика К. М. Быкова [*vignette of valves in veins after Harvey*] Издательство академии наук СССР 1948

22 cm., pp. 234+[2], illustrated.

Note: The translation of *De Motu Cordis* by K. M. Bykov was first published in 1927 (see no. 29 *b*). It is here reprinted with the addition of the same translator's version of *De Circulatione Sanguinis*. These are followed by Pavlov's preface of 1924; two prefaces by Bykov and an account of Harvey's life and works; an oration on Harvey by A. F. Samojlov given at a Harveian celebration in 1928; and, finally, Bykov's commentary and notes.

    5000 copies were printed for the Academy of Sciences at their own press in Leningrad.

Copy: WHML.

29*d*   DE MOTU CORDIS: *DANISH*   4°   1929

Title: William Harvey's Bog Om Opdagelsen af Blodets Kredsløb Oversat og udgivet af V. Meisen København 1929

    22 cm., pp. [xxxii] + 103 + [3], illustrated.

Copy: WHML.

29*e*   DE MOTU CORDIS: *LATIN AND SPANISH*   4°   1936

Title: Harvey Iniciador del Metodo Experimental Estudio critico de su obra De Motu Cordis y de los factores que la mantuvieron ignoranda en los paises de habla Española, con una reproducion facsimilar de la edicion original y su primera version Castellana par Jose Joaquin Izquierdo...
Ediciones Ciencia Mexico 1936

    23 cm., pp. xviii + 398 + [2], illustrated.

Note: A facsimile reproduction of the first edition of *De Motu Cordis* occupies pp. 127–200; the additional leaves K 1–2 are not included. Pp. 267–387 give the first translation of the work into Spanish by J. J. Izquierdo, Professor of Physiology in the University of Mexico.

Copies: BML.
     G. L. Keynes.

29*f*   DE MOTU CORDIS: *SPANISH*   8°   1944

Title: Estudio Anatómico del movimento del corazón y de la sangre en los animales.
Buenos Aires Emecé editores [1944]

    17·3 cm., pp. 192 + [4].

Note: This is a reprint of the translation by Professor Izquierdo first published in 1936 (see no. 29*e*).

Copy: NYAM.

II

DE CIRCULATIONE SANGUINIS

'In my book concerning the motion of the heart and blood in creatures, I only chose out those things out of my many other observations, by which I either thought that errors were confuted, or truth was confirm'd; I left out many things as unnecessary and unprofitable, which notwithstanding are discernable by dissection and sense; of which I shall now add some in few words, in favour of those that desire to learn.' (William Harvey, *Exercitatio altera*, 1649. Nonesuch edition, 1928, p. 147.)

BIBLIOGRAPHICAL PREFACE

Harvey's first treatise caused, as he no doubt expected, a great stir among the philosophers and anatomists of his time. As already seen in the preceding section, several of the reprints of his tract were accompanied, paragraph by paragraph, by the 'refutations' of those who disliked to acknowledge that accepted authority was at fault. Harvey was subjected to misrepresentation and abuse, but in spite of every provocation he maintained silence for twenty-one years, secure in the possession of the truth. At length in 1649 he chose to publish a small book containing two essays addressed to John Riolan the younger, professor of anatomy at Paris. Riolan was one of the most famous anatomists of his time, and so was perhaps regarded by Harvey as specially worthy of an answer, although by no means all his earlier opponents had been lesser lights. Riolan had expressed his views regarding the circulation of the blood in a book entitled: *Opuscula anatomica nova. Quæ nunc primum in lucem prodeunt. Instauratio magna Physicæ & Medicinæ, per novam Doctrinam de Motu Circulatorio Sanguinis in Corde. Londini, Typis Milonis Flesher. MDCXLIX.*

Harvey's book was published in 1649 by Roger Daniel at Cambridge and by Arnold Leers at Rotterdam. Both books are rare, so rare that the very existence of the Cambridge edition was doubted by Mr Charles Perry Fisher when compiling his list in 1912, but now I am able to record twelve copies of one and thirteen of the other. The Cambridge edition has the greater bibliographical interest, as the first title-page was cancelled, and the substituted title-page is found in two forms. No example of the original title-page is known at present, but the copy of the book in the Bodleian Library shows part of its left-hand edge still remaining. Only a few initial capitals can be seen, but the spacing between them is different from the arrangement on the inserted title-page. An example of the book with the first title-page may yet come to light. In the first form of the substituted title a word had been erroneously inserted in the imprint and was erased. This was afterwards corrected, the word being removed and the type closed up, as can be seen in the reproduction given here. The corrected title is usually associated with a second leaf of errata inserted at the end. The Cambridge University Library is, properly enough, the only library possessing both issues of the book. These, which were formerly in the collection of the late librarian, Francis Jenkinson, are both perfect copies in their original bindings, and Jenkinson recorded in a note written in 1919 that the one 'cost 10*d*. originally; but I had to give 3*s*. 6*d*. for it. The other cost 4*s*.'

The Rotterdam edition is a well-printed little book. It is always assumed to have been printed after the Cambridge edition, though I have not obtained any definite proof of this. Its title-page is the more correctly worded, *Exercitationes Duæ Anatomicæ* representing the truth better than the *Exercitatio Anatomica* of the Cambridge edition.

Since its first appearance *De Circulatione Sanguinis* has only once been reprinted separately. This was in the Paris edition of 1650, printed *juxta exemplar Cantabrigiæ*. So rare is this issue, that I have only been able to trace four copies, one being in the library of the Faculté de Médecine at Paris. It was published by Caspar Meturas, who also published Riolan's reply to Harvey in 1652 in a book entitled: *Ioannis Riolani Reſponſio Ad duas Exercitationes Anatomicas poſtremas Guillelmi Harvei De Circulatione Sanguinis. Pariſiis. M.DC.LII.* 12°.

The treatise is not of primary importance and since 1650 it has been properly regarded as merely an appendix to *De Motu Cordis*. It was first published with this in 1654 (no. 8), and the two have usually appeared together since that date, whether in Latin or in translations. The further bibliographical history of *De Circulatione Sanguinis* is therefore almost exactly the same as that of *De Motu Cordis* and need not be recapitulated.

It must be recorded, however, that a new translation of the two letters to Riolan into French by Dr L. Chauvois was published in *Biologie Médicale*, vol. XLII (Paris, 1953), together with a photographic reproduction of the Latin original from the pages of the *Opera Omnia* (Leyden, 1737).

30　DE CIRCULATIONE SANGUINIS　　　12°　1649

Title: Exercitatio Anatomica De Circulatione Sanguinis. Ad Joannem Riolanum filium Parifienfem; medicum peritiffimum Anatomicorũ Coryphæum: in Academia Parifienfi Anatomes & Herbariæ Profefforem Regium & egregium atque Decanum, Reginæ matris Lodovici XIII medicum primarium. Authore, Gulielmo Harveo Anglo, in Collegio Medicorum Londinenfium Anatomes & Chirurgiæ Profeffore; fereniffimæque Majeftati Regio Archiatro. [*rule*] Cantabrigiæ, Ex officina Rogeri Danielis, almæ Academiæ Typographi. 1649.
Proftant venales prope [*word erased*] oftiolum Boreale Templi Divi Pauli, Londini.

Collation: A–E¹², F⁴; 64 leaves.

Contents: A1 title (cancel leaf); A2*a*–F3*b* (pp. 1–124) *De Circulatione Sanguinis*; F4*a Typographus infpectori erudito*; F4*b* blank.

Note: The page number of p. 70 is missing. The pagination is otherwise correct. A word has been erased from the imprint in all the copies known to me. The title-page was subsequently corrected, and a few copies contain this later form with some additional errata inserted at the end (see next entry).

Copies: BM, BWU, CPP, RCP (lacks F4), RCS, ULC, WHML.

31　DE CIRCULATIONE SANGUINIS　　　12°　1649

Title: Exercitatio Anatomica De Circulatione Sanguinis.... [*etc. as in no.* 30]. Cantabrigiæ, Ex officina Rogeri Danielis, almæ Academiæ Typographi. 1649.
Proftant venales prope oftiolum Boreale Templi Divi Pauli, Londini.

Collation: A–E¹², F⁴, with 2ll. inserted; 66 leaves.

Contents: A1 title (cancel leaf); A2–F4 as in no. 30; two leaves inserted at the end, the recto of the first with additional errata: *Hæc errata, quoque utpote graviora, senfum fuum vel pervertentia vel obfcurantia, prius corrigenda calamo, quam legenda, Author confulit.* The verso and the whole of the second leaf is blank.

Note: The book is sometimes found in this form with a corrected title-page and with a sixth of a sheet, carrying additional errata, inserted at the end. It is not to be regarded as a different edition from no. 30, but merely as a later issue.

Copies: BLO, GUL (Hunterian Collection; lacks the second leaf of *errata*), LUM, LUN, ULC.

Exercitatio Anatomica

De

CIRCuLATIONE
SANGUINIS.

Ad *Joannem Riolanum* filium *Parisiensem*; medicum peritissi-mumAnatomicorū Coryphæum: in Academia *Parisiensi* Anatomes & HerbariæProfessorem Regium & egregium atque Decanum, Reginæ matris Lodovici XIII medicum primarium.

Authore,
GULIELMO HARVEO Anglo, in Collegio Medicorum *Londinensium* Anatomes & Chirurgiæ Professore; serenissimæque Majestati Regio Archiatro.

CANTABRIGIÆ,
Ex officina *Rogeri Danielis*, almæ Academiæ Typographi.
1 64 9.
Proſtant venales prope oſtiolumBoreale
Templi Divi Pauli, LONDINI.

Exercitatio Anatomica

De

CIRCuLATIONE
SANGUINIS.

Ad *Joannem Riolanum* filium *Parisiensem*; medicum peritissi-mumAnatomicorū Coryphæum: in Academia *Parisiensi* Anatomes & HerbariæProfessorem Regium & egregium , atque Decanum, Reginæ matris Lodovici XIII medicum primarium.

Authore,
GULIELMO HARVEO Anglo, in Collegio Medicorum *Londinensium* Anatomes & Chirurgiæ Professore; serenissimæque Majestati Regio Archiatro.

CANTABRIGIÆ,
Ex officina *Rogeri Danielis*, almæ Academiæ Typographi.
1 64 9.
Proſtant venales prope oſtiolum Boreale
Templi Divi Pauli, LONDINI.

32 DE CIRCULATIONE SANGUINIS   12° 1649

Title: Exercitationes Duæ Anatomicæ De Circulatione Sanguinis. Ad
Joannem Riolanum filium; Parifienfem Medicum peritiffimum, Anato-
micorum Coryphæum: in Academia Parifienfi Anatomes & Herbariæ
Profefforem Regium egregium atque Decanum, Reginæ matris Lodo-
vici XIII. Medicum primarium. Authore, Gulielmo Harveo Anglo, in
Collegio Medicorum Londinenfium Anatomes & Chirurgiæ Profeffore;
fereniffimæque Majeftatis Regiæ Archiatro.
Roterodami, Ex Officinâ Arnoldi Leers, 1649.

Collation: A–F¹²; 72 leaves.

Contents: A1 title; A2a–F11b (pp. 1–140) *Exercitatio Anatomica Prima, Altera*;
F12a *Errata*; F12b blank.

Note: Sometimes bound up with *De Motu Cordis, Roterodami, Leers,* 1648.

Copies: BM, BML, BWU, CCC, FMP, GUL, LUM (untrimmed), RCPE, SGL,
WHML, YML.
G. L. Keynes, F. C. Pybus.

33 DE CIRCULATIONE SANGUINIS   12° 1650

Title: Exercitatio anatomica de Circulatione Sanguinis. Ad Joannem
Riolanum filium, Parifienfem Medicum peritiffimum, Anatomicorum,
Coryphæum: in Academia Parifienfi Anatomes, & Herbariæ Profef-
forem Regium, & egregium, atque Decanum, Reginæ matris Ludovici
XIII. Medicum primarium. Authore Gulielmo Harveo, Anglo, in
Collegio Medicorum Londinenfium Anatomes, & Chirurgiæ Profeffore;
fereniffimæque Maieftatis, Regio Archiatro. Juxta exemplar Canta-
brigiæ, editum.
Parisiis, Apud Gafpardum Meturas, viâ Iacobæâ, fub figno SS. Trinitatis,
prope Mathurinenfes. [*rule*] M.DC.L.

Collation: A–C¹² D⁶; 42 leaves.

Contents: A1 title; A2a–B2b (pp. 3–28) *Exercitatio anatomica De circulatione sanguinis*;
B3a–D5a (pp. 29–81) *Exercitatio altera*; D5b *Errata*; D6 blank.

Note: There is no copy in the Bibliothèque Nationale.

Copies: BWU, FMP, LUM (the page-number of p. 3 is misprinted 1), RCP.

Exercitationes Duæ
Anatomicæ
DE CIRCVLATIO-
NE SANGUINIS.

Ad JOANNEM RIOLANUM
filium ; *Parifienfem·*Medicum peritiffimum, Anatomicorum Coryphǫum:
in Academia *Parifienfi* Anatomes &
Herbariæ Profefforem Regium egregium atque Decanum , Reginæ matris Lodovici XIII. Medicum primarium.

Authore ,
GULIELMO HARVEO Anglo,
in Collegio Medicorum Londinenfium *Anatomes & Chirurgiæ Pro-*
feffore ; fereniffimæque
Majeftatis Regiæ
Archiatro.

ROTERODAMI,
Ex Officinâ *Arnoldi Leers,* 1649.

III

DE GENERATIONE ANIMALIUM

'Let us then blush in this so ample and so wonderful field of nature to credit other mens traditions only, and thence coine uncertain problemes, to spin out thorney and captious questions. *Nature* her selfe must be our adviser; the path she chalks out must be our walk: for so while we confer with our own eies, and take our rise from meaner things to higher, we shall be at length received into her Closet-secrets.' (William Harvey, *Anatomical Exercitations* (London, 1653), a6*b*.)

BIBLIOGRAPHICAL PREFACE

HARVEY had for a great number of years experimented and recorded his observations on the development of the chick embryo and of other animals. There are many references to the subject in his writings on the heart and circulation of the blood, and he even mentions in his second essay to Riolan matters 'which I will take notice of in the generation of creatures', so that his book, *De Generatione Animalium*, had certainly taken shape at that date (1649). It is evident from references in the sixth Exercise of this work that this interest in the subject of generation had been initiated by his association with the great Fabricius when he was at Padua, 1598–1602. He may even have assisted in the experiments described by Fabricius, whose book *De formato fetu*, was published in 1600. Although the book had been so long in gestation he was reluctant to publish it, and it was only owing to the importunity of his friend, Dr (afterwards Sir) George Ent, that he eventually in 1651 allowed it to be printed. It may be surmised that his reluctance was due partly to the weariness of his age and infirmities, for Ent gained his point by undertaking to see the book through the press, so that the burden and difficulty of deciphering his own crabbed writing was removed from the author's shoulders. Ent reports in his dedication the conversation with Harvey in which he secured his consent to publication, and remarks at the end that 'as our author writes a hand which no one without practice can easily read, I have taken some pains to prevent the printer committing any very grave blunders through this, a point which I observe not to have been sufficiently attended to in the small work of his which lately appeared [no doubt *De Circulatione Sanguinis*, 1649]'. We have already seen evidence of this in the list of errata in the first edition of *De Motu Cordis*.

Harvey has gained so much renown by his earlier work on the heart and circulation that his later contributions to science have been somewhat neglected. The treatise on the generation of animals is written in a series of seventy-two 'exercises', with three additional chapters on parturition, the structure of the uterus, and conception. In Exercise 51 he refers to the theory of 'epigenesis, or the additament of parts budding one out of another', which Thomas Huxley claimed should 'give him an even greater claim to the veneration of posterity than his better known discovery of the circulation of the blood'.

This, however, is an overstatement, and Professor Arthur William Meyer in his valuable and interesting *Analysis of the De Generatione Animalium*,[1] points out that, though Harvey was not a 'preformationist', his conception of epigenesis was very much simpler than that conveyed by the modern use of the term. Nevertheless, Harvey was much ahead of his time in his ideas, and it is claimed by Dr H. P. Bayon that 'with regard to embryology Harvey's name can be rightly placed between those of Fabrizio and Malpighi'.[2]

Harvey's views as expressed in *De Generatione* have been summarized by Dr Joseph Needham[3] as follows:

1. There can be no doubt that the doctrine *ex ovo omnia* was an advance on all preceding thought.

2. He identified definitely and finally the cicatricula on the yolk membrane as the spot from which the embryo originated.

3. He denied the possibility of generation from excrement and from mud, saying that even vermiparous animals had eggs.

4. He discussed the question of metamorphosis (preformation) and epigenesis and decided plainly for the latter, at any rate for the sanguineous animals.

5. He destroyed once for all the Aristotelian (semen-blood) and Epicurean (semen-semen) theories of early embryogeny.

6. He handled the question of growth and differentiation better than any before, anticipating the ideas of the present century.

7. He settled for good the controversy which had lasted for 2200 years as to which part of the egg was nutritive and which was formative by demonstrating the unreality of the distinction.

8. He set his predecessors right on a very large number of detailed points such as the nature of the placenta.

9. He made a great step forward in his theory of foetal respiration, though here he did not consolidate the gain.

10. He affirmed that embryonic organs were active, and that the embryo did not depend on external aid for its principal physiological functions.

[1] Stanford University Press, California; London, Oxford University Press (1936), p. 39.
[2] *Annals of Science*, vol. III (1938), p. 82.
[3] *A History of Embryology* (Cambridge, 1934), p. 129.

Harvey's *De Generatione Animalium*, clearly a work of very great importance, was first printed and published in London in 1651. The first edition is a handsome quarto with a curious allegorical frontispiece representing Jove seated on a pedestal and holding in his hand an egg inscribed *ex ovo omnia* from which are springing animals, insects and plants. The plate is anonymous as regards both designer and engraver, but the name of Richard Gaywood suggests itself as that of a contemporary craftsman who might well be the author of this plate. Gaywood was a pupil of Hollar, the well-known Czech artist who was closely associated with Harvey, both having accompanied Lord Arundel on a diplomatic mission to Vienna in 1636. His name has also arisen in connexion with an etched portrait of Harvey in his old age which has been attributed to Hollar, but which has seemed to me, on the grounds of style and competence, to be more likely to be the work of Gaywood.[1] Furthermore, it has recently been discovered that this plate was done from the life and was intended to be inserted in the first edition of *De Generatione*. This information is derived from a letter from Dr Jasper Needham, F.R.S., to John Evelyn, dated Covent Garden, 5 April 1649, as follows: 'Dr Harvey's picture is etcht by a friend of mine and should have been added to his work, but that resolution altred: however I'l send you a proof with your book that you may bind it up with his book *De Generatione*. I'm sure 'tis exactly like him, for I saw him sit for it.'[2] Both the letter and the book with the etching inserted are still in the Evelyn collection formerly at Wotton and now housed at Christ Church, Oxford. The portrait, reproduced here from an impression in the British Museum, is unflattering, but appears to be an honest and lifelike representation of Harvey in 1649. The attribution of both etchings to Gaywood must remain for the present conjectural.

The first edition of *De Generatione Animalium* is not a very uncommon book, but its important place in the history of science has stimulated the demand, and it now commands a high price in the second-hand book market. It seems that some copies are printed on better paper than others, so that, while many copies are somewhat browned and fragile, others are

---

[1] See my monograph *The Portraiture of William Harvey* (London, Royal College of Surgeons, 1949), p. 16.
[2] First published in a note by me in the *British Medical Journal* (1 July 1950), vol. II, p. 43.

PLATE 5

sometimes seen printed on thick paper which has remained clean and crisp. The most interesting copy at present known is in the possession of Mr F. C. Pybus, F.R.C.S., of Newcastle-upon-Tyne, who obtained it from a London bookseller in 1924. This copy has on the fly-leaf the signature of Harvey's nephew, Sir Eliab Harvey, with the date 1674, and it evidently came to him either at Harvey's death in 1657 or through his father, Eliab Harvey, who died in 1661, for most of the fly-leaves have manuscript notes on Aristotle in Harvey's hand with his monogram WH put against passages which he thought to be of special interest, as was his usual practice. These notes, dealing with the problem of fertilization, were transcribed by Dr Charles Singer, and were printed in Dr Joseph Needham's *Chemical Embryology*, 1931. The London edition of 1651 was followed in the same year by three editions printed in Holland in duodecimo. The publishers responsible were Elzevir, Jansson, and Ravesteyn. It is usually assumed that the Elzevir edition was the first of these; it is certainly the most elegant, and part of the edition carries on the engraved frontispiece the imprint of Pulleyn, the publisher of the quarto edition, no doubt being intended for the English market. The book was reprinted once in Holland in 1662, again at Padua in 1666, and at the Hague in 1680. The Padua edition, published by Frambotti, like the Cadorinus edition of *De Motu Cordis*, 1689, is of considerable rarity. A small volume of excerpts was published at Amsterdam in 1674 with notes by William Langly.

In 1653 an English translation of *De Generatione Animalium* was published by James Young in the same format and year as the English version of *De Motu Cordis*. At the beginning there is a poem addressed to Harvey written by Dr Martin Llewellyn, the author of *Men Miracles*, London, 1646, and it has sometimes been assumed that he was the translator. Llewelyn does not, however, claim the translation as his in the poem, and there is no evidence that he had any hand in it. The translator is, perhaps, more likely to have been Dr George Ent himself, but there is no clue to his identity. The volume has always commanded a good price when perfect, because it contains an engraved portrait of Harvey, by William Faithorne which is of the very first quality. Many copies of the book lack the engraving, and may never have contained it, but no doubt so fine a work of art, which may have been done by Faithorne from the life, has often been removed by those collectors of engraved portraits who would rather mutilate a book than allow a gap to remain in their portfolios. Faithorne's copper-plate has survived to the present time, having been secured by the late Dr Sydney H. Badcock in 1923 after 41 years' search.[1] It had been in the possession of the bookseller Daniel of Mortimer Street, who had 200 prints made, including 100 for Dr F. W. Cock, in 1906. The plate was a delicate one, and soon lost some of the finer details, so original impressions can be readily distinguished from reprints by the state of the print, if not by the paper on which it is printed.

A curious legend has arisen concerning the English edition, the origin of which is not known. Its propagation is chiefly due to W. C. Hazlitt who records in his *Collection and Notes*, vol. II, pp. 270–1: 'It is said that 150 copies were printed, and of them 115 were destroyed by fire.' The myth has been sedulously fostered by booksellers in their catalogues, but it is certainly untrue, as the book is not of any particular rarity.

The Latin text was printed eleven times in all, including its appearances in the collected *Opera*. The original English translation has never been reprinted, except for the last three chapters which were printed in Manchester in 1849 (no. 43 *b*). A new translation by Dr Robert Willis was made for the *Works* published by the Sydenham Society in 1847, and this seems to have satisfied the demand up to the present time. Professor Meyer had an entirely new translation made for the purposes of his *Analysis*, 1936, but this has not been published.

Alexander Ross was the first commentator on Harvey's book to go into print in the second edition of his *Arcana Microcosmi*, 1652. Very few contemporary writers took further notice, and not many of the Harveian Orators during the last 290 years have given any attention to the subject, although the book has undoubtedly deserved a better fate than to be neglected for so long a time. Professor Meyer's *Analysis*, 1936, and Dr Needham's *Chemical Embryology*, 1931, have done something to re-instate Harvey's reputation as an embryologist.

[1] Letter to Sir D'Arcy Power, 21 September 1930.

Gulielmus Harveus
de
Generatione Animalium.

PLATE 6

EXERCITATIONES
DE
Generatione Animalium.

Quibus accedunt quædam
De Partu : de Membranis ac humoribus Uteri :
& de Conceptione.

AUTORE
GUILIELMO HARVEO
Anglo, in Collegio Medicorum *Londi-*
nensium Anatomes & Chirurgiæ Professore.

LONDINI,
Typis Du-Gardianis; impensis *Octaviani*
Pulleyn in Cœmeterio *Paulino.*
M. DC. LI.

34 DE GENERATIONE ANIMALIUM 4° 1651

Title: Exercitationes de Generatione Animalium. Quibus accedunt quædam De Partu: de Membranis ac humoribus Uteri: & de Conceptione. [*rule*] Autore Guilielmo Harveo Anglo, in Collegio Medicorum Londinenſium Anatomes & Chirurgiæ Profeſſore. [*ornament between rules*].

Londini, Typis Du-Gardianis; impenſis Octaviani Pulleyn in Cœmeterio Paulino. M̄.DC.LI.

Collation: [*]⁴ a⁴ B–Z Aa–Sſ⁴; 168 leaves.

Contents: [*]1 blank; [*]2 frontispiece; [*]3 title; [*]4*a*–a4*a* dedication *Præsidi, & Sociis Collegii Medicorum Londinenſium* by *Georgius Ent S.P.D.*; a4*b* blank; B1*a*–C3*b Præfatio*; C4 blank; D1*a*–Sſ3*a* (pp. 1–301) *De Generatione*, etc.; Sſ3*b Menda, lector candide, ſic emenda*; Sſ4 blank.

Frontispiece: A representation of Jove seated on a pedestal and holding in his hands an egg inscribed: *ex ovo omnia*. He has lifted the upper half of the egg with his right hand and animals, insects, and plants are springing out of the lower half. At his side is an eagle with thunderbolts. A pediment below is inscribed: *Gulielmus Harveus/de/ Generatione Animalium*. In the background is a landscape with buildings. The plate-mark measures 21 × 16 cm.

Note: Sign. C2 and Aa3 are misprinted B2 and Aa5. The pagination is correct.

Copies: BM (presentation copy from Dr Ent), BLO (two copies), BML, BWU, CCC (two copies), CPP, FMP, GUL, LUM, MSL, NYAM, RCP, RCPE, RCS (two copies), RFG, RSM, StBH, SGL, ULC, ULE, WHML (four copies), YML.

G. L. Keynes (two copies, one with Dr John Lawson's book-ticket, and MS poem to Harvey on the fly-leaf), F. C. Pybus (with notes by Harvey).

35 DE GENERATIONE ANIMALIUM 12° 1651

Title: Exercitationes de Generatione Animalium. Quibus accedunt quædam De Partu: de Membranis ac humoribus Vteri: & de Conceptione. Autore Guilielmo Harveo Anglo, in Collegio Medicorum Londinenſium Anatomes & Chirurgiæ Profeſſore. [*device*] Amſtelodami, Apud Ludovicum Elzevirium. CIↃ IↃC LI.

Collation: A–Z Aa¹²; 288 leaves.

EXERCITATIONES
DE
GENERATIONE
ANIMALIVM.

Quibus accedunt quædam

De Partu: de Membranis ac humoribus
Vteri: & de Conceptione.

AVTORE

GVLIELMO HARVEO Anglo,
in Collegio Medicorum Londinensium
Anatomes & Chirurgiæ Professore.

AMSTELODAMI,
Apud Ludovicum Elzevirium.
CI Ɔ I Ɔ C LI.

Contents: A1 engraved title; A2 printed title; A3*a*–A7*b* (pp. 5–14) dedication; A8*a*–B6*b* (pp. 15–36) *Præfatio*; B7*a*–Aa8*b* (pp. 37–568) *De Generatione*, etc., A9*a*–A11*a Index Exercitationum* and *Additamentorum*; A11*b*–A12*b* blank.

Engraved title: A reduced copy of the frontispiece of the first edition, with alterations in detail. Jove is seated on a pedestal beneath a classical archway and has the same attributes as before. At the foot of the pedestal on either side is a bird sitting on a basket of eggs. The top corners of the archway are ornamented with eggs in the act of hatching. On the pedestal is inscribed: *Guilielmi/Harvei/Exercitationes/de/ Generatione/Animalium.* Beneath is the imprint: *Londini,/Apud Octavianum Pulleyn./* 1651. The plate-mark measures 11·5 × 6·5 cm. In some copies the imprint has been altered (see next entry).

Copies: BLO, BML, BWU, CCC, NYAM, RCP, ULE (two copies), WHML (four copies, one lacking engraved title).
G. L. Keynes, F. C. Pybus.

36 DE GENERATIONE ANIMALIUM 12° 1651

Title, collation, contents: As in no. *35.*

Engraved title: Printed from the same plate as in no. 35, but with the imprint altered to read: *Amſtelodami,/Apud Ludovicum Elzevirium.* 1651.

Note: This book was printed by Elzevir at Amsterdam and provided with two forms of engraved title-page for the English and continental markets respectively.

Copies: BM, BLO, BWU, FMP, GUL, LUM, NLS, NYAM, RCS (two copies), SGL, ULC, WHML.
G. L. Keynes, F. C. Pybus.

37 DE GENERATIONE ANIMALIUM 12° 1651

Title: Exercitationes De Generatione Animalium. Quibus accedunt quædam De partu: de Membranis ac humoribus Uteri: & de Conceptione. Autore Guilielmo Harveo Anglo, in Collegio Medicorum Londinenſium Anatomes & Chirurgiæ Profeſſore. [*ornament*] Amſtelodami, Apud Joannem Janſſonium. [*rule*] CIƆ IƆ C LI.

Collation:)(⁽¹²⁾)()(⁽⁶ A–R¹² S⁴ T²; 228 leaves.

Contents:)(1 engraved title;)(2 printed title;)(3*a*–)(6*b* dedication;)(7*a*–)()(5*b Præfatio*;)(6 blank, sometimes cancelled; A1*a*–S4*a* (pp. 1–415) *De Generatione*, etc.; S4*b* blank; T1*a*–T2*b Index Exercitationum* and *Additamentorum*.

Engraved title: A reduced copy of the frontispiece of the first edition. Above Jove's head hangs a curtain, and the egg in his hands bears no inscription; otherwise the attributes are as in the first edition. Below is a shield inscribed: *Guilielmi|Harvei| Exercitationes|de Generatione Animalium.|Amſtelodami,|Apud Ioannem Ianſonium.| A°. 1651.* The plate-mark measures 11·3 × 6·3 cm.

Note: P. 253 is numbered 255 and pp. 338–9, 342–3, 346–7, 350–1, 354–5 are numbered 138, 139, etc.

Copies: BM (imperfect), BWU, CPP, FMP, LUM, NYAM, RSM (lacks engraved title), WHML (three copies), YML.

    G. L. Keynes, F. C. Pybus.

38   DE GENERATIONE ANIMALIUM   12°   1651

Engraved title: Exercitationes de Generatione Animalium. Quibus accedunt quædam De Partu: de Membranis ac humoribus Uteri: & de Conceptione. Autore Guilielmo Harveo Anglo, in Collegio Medicorum Londinenſium Anatomes & Chirurgiæ Profeſſore Editio noviſsima a mendis repurgata.

Amſtelædami Apud Ioannem Raveſteynium. A°. 1651.

The engraving is a partial copy of those in the Elzevir and Jansson editions of the same year. Jove is seated on a pedestal with a curtain overhanging him on his right and with three pillars beyond him on his left. Between the pillars is seen a distant landscape and a crocodile. On the pedestal is inscribed the title given above and the publisher's imprint is at the bottom of the plate. On the ground on either side of the pedestal are birds on baskets of eggs with numerous hatching chicks and reptiles and a tortoise scattered beyond them. The plate-mark measures 11·3 × 6·3 cm.

Collation: (*)¹² (*)(*)² A–Q¹² R⁴; 210 leaves.

Contents: (*)1 engraved title; (*)2*a*–(*)5*a* dedication; (*)5*b*–(*)(*)2*b Præfatio*; A1*a*–R2*a* (pp. 1–388) *De Generatione*, etc.; R3*a*–R4*a Index*; R4*b* blank.

Copies: BM, CCC (with autograph of Prof. T. H. Huxley), CPP, LUM, NYAM, SGL, WHML, YML.

    G. L. Keynes, F. C. Pybus.

39   DE GENERATIONE ANIMALIUM   12°   1662

Engraved title: As in no. 38 with the date altered to 1662.

Collation: As in no. 38.

Contents: (*)1 engraved title; (*)2*a*–(*)4*b* dedication; (*)5*a*–(*)(*)2*b Præfatio*; A1*a*–R2*a* (pp. 1–388) *De Generatione*, etc.; R3*a*–R4*a Index*; R4*b* blank.

Note: A reprint of the edition of Ravesteynius of 1651, using the same plate for the engraved title.

Copies: BM, BWU, CPP, GUL, LUM, MSL, NYAM, RCS, SGL, ULC, WHML, YML.
    G. L. Keynes, F. C. Pybus.

40   DE GENERATIONE   12° 1666

Title: Exercitationes De Generatione Animalium. Quibus accedunt quædam De Partu: de Membranis ac humoribus Vteri: & de Conceptione. Autore Guilielmo Harveo Anglo, in Collegio Medicorum Londinenſium Anatomes & Chirurgiæ Profeſſore. Cum Elencho Exercitationum. [*device*]
Patavii, M.DC.LXVI. [*rule*] Typis Heredum Pauli Frambotti, Bibliop. Superiorum Permiſsu.

Collation: †¹² ††⁶ A–Z Aa,Bb¹² Cc⁶; 324 leaves.

Contents: †1 engraved title; †2 title; †3*a*–†4*a D. Petro Angelo Diamanti Dedicatio* by *Petrus Maria Frambottus*; †4*b* blank; †5*a*–††6*a Præfatio*; ††6*b* blank; A1*a*–Cc2*b* (pp. 1–604) *De Generatione,* &c.; Cc3*a*–Cc5*a Elenchus*; Cc5*b*–Cc6*b* blank.

Engraved title: A copy of the plate in the Amsterdam edition of 1651 (no. 35). The shield below is inscribed: *Guilielmi*|*Harvei*|*Exercitationes*|*de*|*Generatione Animalium.*| *Patavii,*|*Ex Officina Heredum Pauli Frambotti*|*An.° salutis Reparatę 1666.*|*de licᵃ Superiorŭ.*| The plate is signed: *I. Ruphanus f.* The plate-mark measures 11·5 × 6·2 cm.

Note: The pagination and signatures in sheet D are disordered. The pagination is erratic throughout. A copy on 'large paper' was offered by Lauria, Paris, in April, 1951.

Copies: BWU, CPP, MSL, NYAM, RCP, StBH, SGL, WHML, YML (untrimmed).
    G. L. Keynes, F. C. Pybus.

41   OBSERVATIONES EXCERPTÆ   12° 1674

Title: Obſervationes et Hiſtoriæ Omnes & ſingulæ è Guiljelmi Harvei libello De Generatione Animalium excerptæ, & in accuratiſſimum ordinem redactæ. Item Wilhelmi Langly De Generatione Animalium Obſervationes quædam. Accedunt Ovi fæcundi ſingulis ab incubatione diebus factæ Inſpectiones; ut et Obſervationum Anatomico-Med.

Decades quatuor; denique Cadavera Balſamo condiendi modus. Studio
Juſti Schraderi, M.D. [*ornament*]
Amſtelodami. [*rule*] Typis Abrahami Wolfgang, Anno 1674.

Collation: *¹² **⁶ A–K¹²; 138 leaves.

Contents: *1 engraved title; *2a printed title; *2b quotation from *Ecclefiaſt.*, xi, 5;
*3a–*5a *Dedicatoria Dn. Matthæo Slado & Dn. Joh. Swammerdam*; *5b–**4b
Præfatio; **5a latin lines *Ad Dn. I. Schraderum* by *M. Sladus, Med.D.*; **5b
Syllabus Capitum Obſervationum Harvei; **6 blank; A1a–F8a (pp. 1–135) *Ob-
ſervationes ex Harvei libello excerptæ*; F8b–H7a (pp. 136–181) *Wilhelmi Langly
Obſervationes*; H7b–K9a (pp. 182–233) *Obſervationum Anatomico-Medicarum
Decades quatuor*; K9b–K10a (pp. 234–5) *Index Obſervationum*; K10b–K12b
(pp. 236–240) *Condit. Cadaverum.*

Illustrations: (1) Engraved title on *1a; an allegorical group representing scientific
pursuits. On an obelisk is inscribed: *Obſervationes de Generat. Animalium et Anat.
Med.* The plate is signed below: *Romyn de Hooghe ſ*, with the imprint: *Apud Abr.
Wolfganck.* The plate-mark measures 12 × 7 cm. (2–9) Engraved plates representing developing chicks (figs. i–vi) and hernial protrusions (figs. vii, viii) inserted
opposite pp. 137, 138, 143, 144, 145, 180, 208, 209.

Note: Six of the plates are the first illustrations of Harvey's text to be published.

Copies: BM, BLO, BWU, GUL, NYAM, RCP, RCPE, RCS, RSM, WHML
(four copies, two lacking engraved title).

42 DE GENERATIONE ANIMALIUM   12° 1680

Engraved title: A copy of the plate in no. 38. On the pedestal is the title:
Exercitationes de Generatione Animalium. Quibus accedunt quædam
De Partu de Membranis ac humoribus Uteri: & de Conceptione.
Autore Guilielmo Harveo Anglo, in Collegio Medicorum Londinen-
ſium Anatomes & Chirurgiæ Profeſſore Editio noviſſima a mendis
repurgata
Below is inscribed: Hagæ Comitis. Apud Arnoldum Leers. A°. 1680.

Collation: *¹² **⁶ A–Z¹² Aa¹² Bb⁶; 312 leaves.

Contents: *1 engraved title; *2a–*5b dedication; *6a–**6b *Præfatio*; A1a–Bb3b
(pp. 1–582) *De Generatione*, etc.; Bb4a–Bb5b *Index*; Bb6 blank.

Copies: BM, BLO, BML, BWU, CPP, FMP, NYAM, RCP (two copies), RFG,
SGL, WHML, YML.

G. L. Keynes, F. C. Pybus.

DE GENERATIONE ANIMALIUM

In: Harvey's *Opera*, Geneva, 1685; Geneva, 1699; Leyden, 1737; London, 1766. (See nos. 44–7.)

43   DE GENERATIONE ANIMALIUM: *ENGLISH* 8° 1653

Title: Anatomical Exercitations, Concerning the Generation Of Living Creatures: To which are added Particular Difcourfes, of Births, and of Conceptions, &c. By William Harvey, Doctor of Phyfick, and Profeffor of Anatomy, and Chirurgery, in the Colledge of Phyfitians of London. [*ornaments between rules*]
London, Printed by James Young, for Octavian Pulleyn, and are to be fold at his Shop at the Sign of the Rofe in St. Pauls Church-yard. 1653.

Collation: A⁸ a⁸ ¶⁸ B–Z Aa–Nn⁸; 304 leaves.

Contents: A1 blank; A2 title; A3*a*–a1*a* *The Epiftle Dedicatory*; a1*b*–a4*a* lines to Dr Harvey by M.LL., M.D.; a4*b*–¶8*b* *The Preface*; B1*a*–Nn6*b* (pp. 1–566) *Of Generation*; Nn7*a* *Errata*; Nn7*b*–Nn8*b* blank.

Frontispiece: Engraving by Faithorne of a bust of the author. The bust stands on a pediment on which is inscribed: *Guilielmus Harveus.* The plate is signed below: *W: F: fec:* The plate-mark measures 13·5 × 9 cm. (See reproduction.)

Note: The lines to Harvey are by Martin Llewelyn, M.D. Oxon., author of *Men Miracles, with other Poems,* London, 1646. This translation of *De Generatione* has never been reprinted as a whole.

Copies: BM, BLO (portrait added by Douce), BML, BWU, CCC (two copies), CPP, GUL, LUM, NYAM, OL, RCP, RCS (two copies, one lacking the portrait), RFG, RSM (four copies, two with portrait), SGL, ULE (lacking the portrait), WHML (two copies, portrait mounted in one), YML (two copies, one lacking the portrait).
   G. L. Keynes (portrait inserted), F. C. Pybus.

DE GENERATIONE ANIMALIUM: *ENGLISH* 8° 1847

In: The Works of William Harvey, M.D....translated from the Latin... by Robert Willis, M.D....London...MDCCCXLVII. (See no. 48.)

GVLIELMVS HARVEVS

PLATE 7

ANATOMICAL EXERCITATIONS,

Concerning the
GENERATION
Of Living Creatures:

To which are added Particular Difcourfes,
of *Births,* and of *Conceptions,* &c.

By *WILLIAM HARVEY,* Doctor
of *Phyfick,* and Profeffor of *Anatomy,*
and *Chirurgery,* in the COLLEDGE
of Phyfitians of *LONDON.*

LONDON,

Printed by *James Young,* for *Octavian
Pulleyn,* and are to be fold at his Shop at the
Sign of the Rofe in St. *Pauls* Church-
yard. 1653.

43*a*   DE GENERATIONE ANIMALIUM: EXCERPTS:
    ENGLISH   8°   1849

Title: On Birth and Conception. By William Harvey, M.D. Reprinted
from the translation of the celebrated Dr Ent, as a monograph for the
British Record of Obstetric Medicine and Surgery, Edited by Charles
Clay, M.D....Manchester: William Irwin, 39, Oldham Street...and
Henry Renshaw, 356, Strand. [London]
22 cm., pp. 58.

Note: Included in *The British Record of Obstetric Medicine and Surgery for 1849...
Vol. II, Manchester:*...1849. The last three chapters of *De Generatione* are reprinted from the English edition of 1653. The editor, Dr Clay, assumes that this
translation was the work of Dr Ent, who, he says, was Harvey's 'literary assistant'.
He further states that 'Dr Ent accompanied Harvey to Amsterdam to correct the
sheets of his Latin edition published in 1651', but gives no authority for this most
improbable event.

Copies: BM, BML, RCS, &c.

IV
OPERA OMNIA

Dii laboribus omnia vendunt
(Harvey's motto in an album, 1641)

BIBLIOGRAPHICAL PREFACE

HARVEY'S chief works in Latin have only twice been printed in a collected form, first by van Kerckhem at Leyden in 1737, and secondly by Bowyer for the Royal College of Physicians in 1766. The latter is an imposing volume with a fine engraved portrait, and worthily enshrines the pious memory of the author. The works had also been printed towards the end of the seventeenth century in two editions of the *Bibliotheca Anatomica* of Le Clerc and Manget. The volume published by the Sydenham Society in 1847 contains the only collected edition of Harvey's works in English. These were translated by Dr Robert Willis and have already been referred to in the prefaces to the preceding sections. The volumes of 1766 and 1847 were also furnished with lives of the author, and certain minor writings were added at the end.

44 BIBLIOTHECA ANATOMICA: OPERA   f° 1685

Title: Bibliotheca Anatomica five recens in Anatomia Inventorum Thes-
aurus...Argumenta, Notulas, & Obfervationes Anatomico-Practicas
addiderunt Daniel Le Clerc & I. Iacobus Mangetus, M.M. D.D....
Genevæ, fumptibus Joannis Anthonii Chouët. M. DC. LXXXV.

Collation: 2 vols. f° in sixes.
Vol. I. Ddd1*a*–Ooo6*b* (pp. 595–728) *Exercitationes de Generatione Animalium.*
Vol. II. D1*a*–G4*a* (pp. 37–79) *De Motu Cordis et Sanguinis, De Circulatione
Sanguinis.*

Note: The figures are on plate 39.

Copies: BML, CPP, OL (lacking plate 39), RCP, WHML (five copies).

45 BIBLIOTHECA ANATOMICA: OPERA   f° 1699

Title: Bibliotheca Anatomica five recens in Anatomia Inventorum Thes-
aurus,...Argumenta, Notas, & Obfervationes Anatomico-Practicas
addiderunt, Daniel Clericus & J. Jacob Mangetus MM, D.D. Editio
secunda....
Genevæ; fumptibus Johan Anthon. Chouët & Davidis Ritter. M. DC. XCIX.

Collation: 2 vols. f° in sixes.
Vol. I. Mm2*b*–Zz2*b* (pp. 412–544) *Exercitationes de Generatione Animalium.*
Aaaa6*a*–Eeee2*b* (pp. 839–880) *Exercitatio de Motu Cordis. Exercitationes de
Circulatione Sanguinis.*

Note: The figures are on plate 42.

Copies: BM, BML, BWU, RCS, RSM, WHML (three copies).

46 OPERA   4° 1737

First title: Exercitatio Anatomica De Motu Cordis et Sanguinis in Ani-
malibus. Cui accedunt Exercitationes Duæ Anatomicæ De Circulatione
Sanguinis Ad Joannem Riolanum Filium; In Academia Parifienfi
Anatomes & Herbariæ Profefforem Regium, Reginæ Matris Ludovici
XIII. Medicum Primarium. Auctore Gulielmo Harveo Anglo, Ana-
tomiæ & Chirurgiæ in Collegio Medic. Lond. Profeffore, Sereniffi-
mæque Majeftatis Regiæ Archiatro. Hujusque Operum Pars Prima.
Editio Noviffima. Indice ornata. [*engraved vignette*]
Lugduni Batavorum, Apud Johannem van Kerckhem, 1737.

Second title: Exercitationes de Generatione Animalium. Quibus accedunt quædam De Partu: De Membranis ac Humoribus Uteri: et De Conceptione. Auctore Gulielmo Harveo, Anglo, in Collegii Medicorum Londinenſium Anatomes & Chirurgiæ Profeſſore; Regis Archiatro. Illiusque Operum Pars Altera. Editio Noviſſima. Indice ornata. [*engraved vignette*] Lugduni Batavorum, Apud Johannem van Kerckhem, 1737.

Collation: [a]² *⁴ (first leaf cancelled) **⁴ A–X⁴ Y², *–***⁴ A–Z Aa–Zz Aaa–Iii⁴ Kkk²; 330 leaves.

Contents: FIRST PART: *1 cancelled; [a]1 half-title; [a]2 first title; *2*a*–**3*a Bernardi Siegfried Albini Praefatio ad medicinae studiosos*; **3*b*–4*b* blank; A1*a*–N4*a* (pp. 1–103) *De Motu Cordis* with the dedications; N4*b* blank; O1 sub-title to *Exercitationes Duæ de Circulatione Sanguinis*; O2*a*–X4*a* (pp. 107–167) *Exercitationes Duæ*; X4*b* blank; Y1*a–b* (pp. 169–170) *Libri Quos excudit...J. A. Kerckhem*; Y2 blank (as in no. 14).

SECOND PART: *1 half-title; *2 second title; *3*a*–*4*b Dedicatio*; **1*a*–***3*b Præfationes*; ***4*a–b Index*; A1*a*–Eee2*b* (pp. 1–404) *De Generatione*; Eee5*a*–Kkk1*b Index rerum notabilium*; Kkk2 blank.

Illustrations: A large folding plate with copies of the usual figures is inserted between K4 and L1 in part 1. The plate is signed *J. d. Groot del J. vd. Spyk fecit.* The plate-mark measures 16·5 × 26·5 cm.

Note: The first part had already been published separately in 1736 (see no. 14). The half-sheet [*a*] carrying the half-title and first title of the present collection has therefore been substituted for the title-page of 1736, which has been cancelled. Y2 blank may also have been cancelled by the binder.

A variant issue in the Cushing collection (YML) has the first title in the same setting of type except that the engraved vignette is omitted and an impression of a printer's block of suitable size pasted in its place. The rest of the first sheet has been reset. There is no half-title, and the collation therefore runs *⁴ **⁴ etc.

Copies: BML, BWU, CPP, LUM, NYAM, RCP, RCPE, RSM, SGL, ULE (two copies), WHML, YML.

47 OPERA   4° 1766

Title: Guilielmi Harveii Opera Omnia: A Collegio Medicorum Londinensi Edita: MDCCLXVI.

Colophon: Londini, Excudebat Guilielmus Bowyer, MDCCLXVI.

Collation: [*]⁴ a–e⁴ A–Z Aa–Zz Aaa–Zzz 4A–4L⁴ 4M² 4N–4Q⁴; 370 leaves.

Contents: [*]1 title; *2a–b *Præfatio*; *3a–*4b *Argumenta Capitum*; a1a–e3b (pp. i–xxxviii) *Guilielmi Harveii Vita*; e4a *emendanda*; e4b blank; A1a–Qq4b (pp. 1–312) text; Rr1 half-title to *Tomus Secundus*; Rr2a–4M2b (pp. 313–642) text; 4N1a–4Q4a *Index rerum*; 4Q4b blank. *Emendanda* and colophon on 4Q4a.

Contents:
Guilielmi Harveii Vita.
Exercitatio de Motu Cordis.
Exercitationes Duæ de Circulatione Sanguinis.
Exercitationes de Generatione Animalium.
Anatomia Thomæ Parri.
Epistolæ.
Diploma Patavinum Harveio concessum.

Illustrations: i. Frontispiece. Portrait of Harvey inscribed below: *Guilielmus Harveius, Colleg. Medicor. Londin. Socius. E pictura Archetypa in Ædibus Collegii Medicorum Londinensis aſservata. Corn͵ Ionson pinxt. I. Hall sculp. Londini.* The plate-mark measures 25 × 18 cm.
ii. Facing p. 68. A plate with the usual figures, signed: *J. Mynde Sc.* The plate-mark measures 23·5 × 17·5 cm.

Note: The text of this edition was very carefully prepared and at the end of the section containing the writings on the heart is a list of 246 emendations in the Frankfort edition of *De Motu Cordis,* 1628, and of 158 in the Cambridge edition of *De Circulatione Sanguinis,* 1649. The volume was edited by Mark Akenside, M.D., and the memoir was by Thomas Lawrence, M.D.

Copies: BML, BWU, CCC, CPP, MSL, NLS, NYAM, RCP, RCPE, RCS, RSM (copy given to Richard Owen at the unveiling of Harvey's statue at Folkestone), StBH, SGL, ULE, WHML (three copies; one with inscription by Dr Akenside; another lacks portrait and title-page).
F. C. Pybus.

48 WORKS   8° 1847

Title: The Works of William Harvey, M.D. . . . Translated from the Latin with A Life of the Author by Robert Willis, M.D. . . .
London Printed for the Sydenham Society MDCCCXLVII.

Collation: 22·5 cm., pp. xcvi + 624, in eights.

Contents:

The Life of William Harvey.
The Will of Harvey.
An Anatomical Disquisition on the Motion of the Heart and Blood in Animals.
An Anatomical Disquisition...to John Riolan, Jun....
A Second Disquisition to John Riolan, Jun.
Anatomical Exercises on the Generation of Animals.
The Anatomical Examination of the Body of Thomas Parr.
Letters.

Copies: BM, BML, BWU, CPP, RCS, StBH, WHML, &c.

V

MISCELLANEA

'I pray pardon this scribling on the grass in the feild and procure with all expedition my freedom from this barbarous usadg.' (In a letter to Lord Feilding in Venice from William Harvey at Treviso, August 3, 1636.)

BIBLIOGRAPHICAL PREFACE

HARVEY's Miscellaneous Writings are not extensive. Only one piece by him has yet been discovered to have been printed during the seventeenth century. This was his account of the post-mortem examination of the body of Thomas Parr who died in 1635 at the reputed age of nearly 153 years. The MS. was given by Michael Harvey, a nephew of the author, to Dr John Betts, with whose writings it was printed in 1669. Most of Harvey's remaining MSS. are believed to have perished when the house of the College of Physicians was burnt in the Fire of London, 1666. Some, however, survived, and came into the possession of Sir Hans Sloane, with whose collections they were bought for the British Museum in 1754. By far the most important of these are his *Prelectiones*, or notes for his lectures on the circulation, dated 1616. This MS. has been reproduced in facsimile (no. 52) and is now well known. Extracts from another Sloane MS., *de Musculis, motu locali, &c.*, were printed by Dr G. E. Paget in 1850 (no. 50). The remainder of the Miscellaneous Writings include his Will, a motto in an album, and letters, of which only fourteen have survived in the original MSS. Twelve of these are in the library of the Royal College of Physicians; the other two are in the Bodleian Library, and the library of Sidney Sussex College.

49 NARRATIO ANATOMICA   8° 1669

Title: De Ortu et Natura Sanguinis [*rule*] A Joanne Betto M.D. Regis
   Medico Ordinario & Collegii Londinenſis Socio. [*quotation between rules*]
   Londini, Ex Officinâ E. T. væneuntque apud Gulielmum Grantham, ad
   Inſigne Urſi nigri, in Aulâ Weſtmonaſterienſi. 1669.

Collation: A⁸ a⁸ b⁴ B–X⁸ Y⁴; 192 leaves.

Contents: A1 blank; A2 title: A3*a*–a7*a* *Ad Lectorem*; a7*b* blank; b1*a*–b4*a* *Elenchus
   Capitum & Sectionum*; b4*b* blank; B1*a*–X6*b* (pp. 1–316) *De Ortu et Natura San-
   guinis*; X7 sub-title to *Anatomia Thomæ Parri Annum Centeſimum quinquageſimum
   ſecundum & novem menſes agentis. Cum Cl. Viri Guliellmi Harvæi Aliorumque
   Adſtantium Medicorum Regiorum Obſervationibus*; X8*a*–Y3*a* (pp. 319–325)
   Narratio Anatomica; Y3*b* blank; Y4*a* *Errata*; Y4*b* blank.

Note: An abstract of this *Anatomical Account* was printed in the *Philosophical Transactions* of the Royal Society, vol. III, 1669, pp. 886–8. It was first reprinted in full
   in the *Opera Omnia*, 1766 (see no. 47). It was translated into English by Dr Robert
   Willis and printed with the *Works* in 1847 (see no. 48). It was next translated
   into German and published with annotations in Carl Mettenheimer's *Sectiones
   Longævorum*, Frankfort, 1863, 8°, pp. 19–25. A new English translation was
   made by Arnold Muirhead and was published, with a note by Geoffrey Keynes,
   in *St Bartholomew's Hospital Reports*, vol. LXXII, 1939.

Copies: BM, BWU, LUM, RCP, RCS, WHML, &c.
       G. L. Keynes, F. C. Pybus.

50 DE MUSCULIS, MOTU, ETC.   8° 1850

Notice of an unpublished manuscript of Harvey. By G. E. Paget, M.D....
London: Longman, Brown, Green and Longmans.
Cambridge: J. Deighton. M. DCCC. L. 21·5 cm., pp. 20.

Note: Contains a letter from Harvey to Samuel Ward, Master of Sidney Sussex College,
   Cambridge, and extracts from his MS. *de Musculis, motu locali, &c.*, 1627, which
   is among the Sloane MSS. in the British Museum. The letter was also printed in
   facsimile in a pamphlet by Dr Paget dated Caius College, 25 October 1849. The
   pamphlet, a copy of which is in the British Museum, is an oblong octavo of four
   leaves. The letter to Ward was first recorded by Zacharias Conrad von Uffenbach
   in 1710 (see *Cambridge under Queen Anne*, ed. Mayor, Cambridge, 1911), and it
   lay beside the *cranium petrifactum* which is the subject of the letter. The *cranium*
   is still in its cabinet, but the letter is now preserved separately in a glazed frame.
   It was last printed by Sir D'Arcy Power in his *Life of Harvey*, 1897.

Copies: BM, BWU, RCS, RSM, &c.

51 MISCELLANEA   8° 1875

Memorials of Harvey. Including a letter and autographs in facsimile.
Collected and edited by J. H. Aveling, M.D.
London: J. & A. Churchill,...1875.

    21 cm., pp. 27+1, with lithographic facsimile of letter inserted as
frontispiece.

Note: Includes the following writings by Harvey: A letter probably to Lord Dorchester,
    c. 1631 (Bodleian Library, Clarendon papers, 2076), Two Certificates, Report on
    the Lancashire Witches, Motto in Album, Report on the Health of Prince Maurice.

Copies: BM, RSM, WHML, &c.

52 PRELECTIONES   4° 1886

Title: Prelectiones Anatomiæ Universalis by William Harvey Edited
    with an autotype reproduction of the original by a committee of the
    Royal College of Physicians of London.
    London J. & A. Churchill 11 New Burlington Street 1886

Collation: A–Z AA–BB⁴ CC²; 102 leaves.

Contents: A1 title; A2*a*–A4*b* (pp. iii–viii) Introduction; B1*a*–CC2*b* transcript of the
    MS. Inserted between each pair of leaves is a leaf with autotype reproductions of the
    pages of the MS.; 98 leaves are inserted.

Note: This MS., now in the Sloane collection at the British Museum, contains notes
    in Harvey's hand of the lectures in which he first described his observations on the
    circulation of the blood. The title-page of the MS. is dated 1616.

    It is recorded in the Minutes of the Council of the Royal College of Surgeons,
    9 December 1886, that 300 copies were printed, of which each Royal College took
    100 at the subscription price of two guineas. Copies are still in stock.

Copies: BM, BWU, RCP, RCS, RSM, StBH, WHML, YML, &c.

53 LETTERS   8° 1912

Some recently discovered letters of William Harvey with other miscellanea
    By S. Weir Mitchell, M.D., LL.D., F.R.S....With a bibliography
    of Harvey's works By Charles Perry Fisher...Philadelphia 1912
    [Transactions of the College of Physicians of Philadelphia].

    23 cm., pp. 59+[3], with two illustrations of Harvey's coffin inserted.

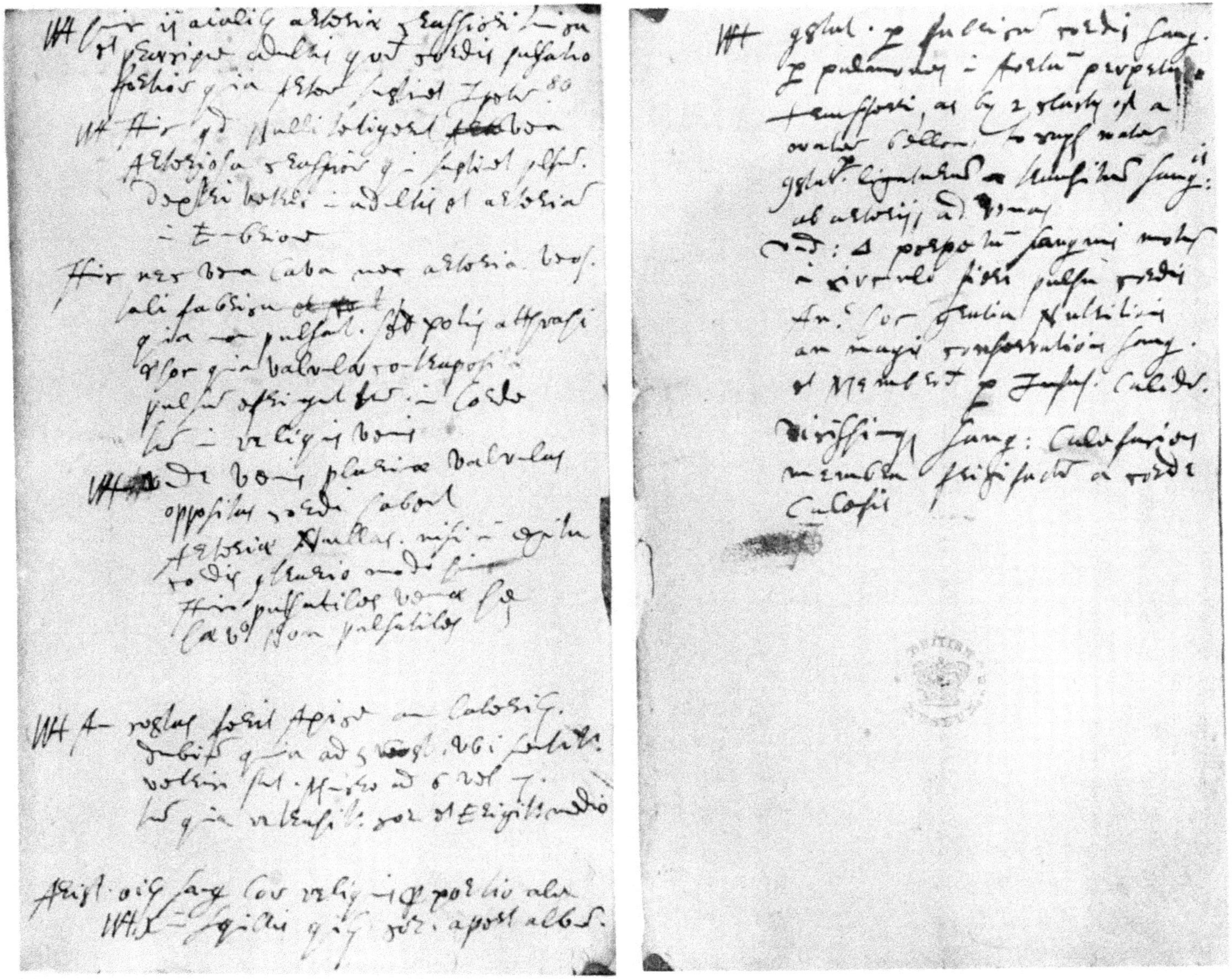

8. HARVEY'S PRELECTIONES 1616

Note: These letters were first printed in the *Hist. MSS. Comm. Report of the MSS. of the Earl of Denbigh*, 1911, pt. v, pp. 28–41, with note at pp. ix–x. They were also printed in a contribution by Sir D'Arcy Power to the *Proceedings* of the Historical Section of the Royal Society of Medicine (1916, x, pp. 33–59), and in a private edition (see next entry).

Copies: BM, BWU, RCP, RCS, RSM, WHML, YML, &c.
    G. L. Keynes.

54 LETTERS   8°   1912

Eleven Letters of William Harvey to Lord Feilding. June 9–Nov. 15, 1636. Purchased from the Earl of Denbigh and presented to the Royal College of Physicians by Sir Thomas Barlow, Bt. President. 18 October, 1912.

22 cm., pp. [30].

Note: A small edition was privately printed for Sir Thomas Barlow, Bt. There is a preliminary note by Francis Jenkinson, Cambridge University Librarian.

COPIES RECORDED
PRINTERS, BOOKSELLERS, & PUBLISHERS
GENERAL INDEX

INDEX OF COPIES RECORDED

(The numbers refer to the entries in the Bibliography)

Aberdeen, King's College Library, 1

Baltimore, Johns Hopkins Medical School Library, 1
Bodmer, Dr Martin, 1
Bonn, University Library, 1
Boston Medical Library [BML], 1, 3, 5, 6, 7, 8, 9, 16, 19, 20, 21, 22, 23, 27, 29, 29e, 32, 34, 35, 42, 43, 43a, 44, 45, 46, 47, 48
Breslau, University Library, 1
Buffalo, Museum of Science, 1

Cambridge, Caius College [CCC], 1, 6, 7, 9, 32, 34, 35, 38, 43, 47
Cambridge, Trinity College, 1
Cambridge University Library [ULC], 3, 7, 9, 10, 11, 23, 30, 31, 34, 36, 39
Clendinning, Dr Logan, 1
Columbia University Library, 1

Edinburgh, National Library of Scotland [NLS], 16, 36, 47
Edinburgh, Royal College of Physicians [RCPE], 8, 15, 20, 32, 34, 41, 46, 47
Edinburgh, University Library [ULE], 1, 5, 7, 8, 16, 20, 34, 35, 43, 46, 47
Erlangen University Library, 1

Glasgow, Royal Faculty [RFG], 3, 9, 15, 19, 20, 34, 42, 43
Glasgow University Library [GUL], 1, 3, 4, 10, 11, 15, 31, 32, 34, 36, 39, 41, 43
Göttingen, University Library, 1

Jena, University Library, 1

Keynes, G. L., 3, 4, 7, 9, 10, 11, 12, 13, 15, 16, 19, 20, 29e, 32, 34, 35, 36, 37, 38, 39, 40, 42, 43, 49

Levy, Dr R., 1
London, British Museum [BM], 1, 2, 3, 4, 5, 6, 7, 8, 9, 11, 15, 16, 19, 20, 21, 22, 23, 26, 30, 32, 34, 36, 37, 38, 39, 41, 42, 43, 43a, 45, 48, 49, 50, 51, 52, 53

London, Medical Society of London [MSL], 4, 7, 9, 11, 15, 20, 34, 39, 40, 41, 47
London, Royal College of Physicians [RCP], 1, 3, 4, 7, 8, 9, 10, 11, 12, 15, 16, 19, 20, 28, 29, 30, 33, 34, 35, 40, 41, 42, 43, 44, 46, 47, 49, 52, 53
London, Royal College of Surgeons [RCS], 1, 3, 5, 8, 10, 11, 19, 20, 21, 23, 25b, 28, 30, 34, 36, 39, 41, 43, 43a, 45, 47, 48, 49, 50, 52, 53
London, Royal Society of Medicine [RSM], 1, 6, 7, 8, 15, 16, 19, 20, 21, 22, 28, 34, 37, 43, 45, 46, 47, 50, 51, 52, 53
London, St Bartholomew's Hospital [StBH], 8, 25a, 34, 40, 47, 48, 52
London, Wellcome Historical Medical Library [WHML], 1, 3, 5, 6, 7, 9, 10, 11, 13, 15, 16, 19, 20, 21, 22, 23, 27, 28, 29c, 29d, 30, 32, 34, 35, 36, 37, 38, 39, 40, 41, 42, 43, 44, 45, 46, 47, 48, 49, 51, 52, 53

McGraw, Dr, 1
Michigan, Library of the University [LUM], 1, 3, 4, 7, 9, 15, 20, 25b, 31, 32, 33, 34, 36, 37, 38, 39, 43, 46, 49
Montreal, McGill University Medical Library [MGU], 14
Montreal, Osler Library, McGill University [OL], 1, 3, 12, 15, 16, 18, 19, 25b, 27, 43, 44
Munster, University Library, 1

Nebraska, Library of the University [LUN], 31
New Haven, Yale Medical Library [YML], 1, 3, 4, 8, 9, 11, 12, 14, 15, 19, 20, 27, 32, 34, 37, 38, 39, 40, 42, 43, 46, 52, 53
New Haven, Yale University Library, 1
New York, Academy of Medicine [NYAM], 1, 3, 4, 5, 8, 9, 10, 13, 15, 16, 19, 20, 23, 27a, 28, 29f, 34, 35, 36, 37, 38, 39, 40, 41, 42, 43, 46, 47

Oxford, Bodleian Library [BLO], 1, 6, 7, 15, 20, 23, 31, 34, 35, 36, 41, 42, 43

PRINTERS, BOOKSELLERS AND PUBLISHERS
1628–1952

(The numbers refer to the entries in the Bibliography)

Academy of Sciences, Leningrad, 29*c*

Barth, J. A., 27, 27*a*
Baxter, D., 15
Bell, George, and Sons, 22
Blaev, Johannes, 5
Bowyer, William, 47
Brewster, John, 16
Brogiollus, M. A., 2

Cadorinus, 12
Carfrae, John, 16
Chouët, J. A., 44, 45
Churchill, J. and A., 51, 52
Ciencia, Editiones, 29*e*

Daniel, Roger, 10, 30, 31
Deighton, J., 50
Dent, J. M. and Sons, 24
Doin, G. et Cie, 29*a*
Dugard, William, 34
Dutton, E. P., and Co., 24

Elzevir, Ludovic, 35, 36
Emecé, 29*f*
Enke, Ferdinand, 26
Enschedé Press, 25

Fitzer, William, 1
Frambotti, Pauli, Heredes, 40

Gilliflower, Math., 20
Grantham, William, 49

Irwin, William, 43*a*

Jansson, John, 37

Kerckhem, John van, 14, 46

Last, Cornelis, 18
Leach, Francis, 19
Leers, Arnold, 7, 8, 9, 11, 32, 42
Lier, R. and Co., 17, 17*a*
Longhus, 13
Longman and Co., 16, 50
Lowndes, Richard, 19, 20

Maire, John, 3, 6
Masson, G., 28, 29
Meisen, V., 29*d*
Meturas, Caspar, 33
Moreton, G., 23

Nonesuch Press, 25

Pecker, Cornelius de, 14*a*
Potuliet, Gerard, 14*a*
Pulleyn, Octavian, 34, 35, 43

Ravesteyn, John, 38, 39
Renshaw, Henry, 43*a*
Renshaw and Rush, 21
Ricciard, Dominic, 4
Ritter, David, 45
Roest, Adrian, 18

Sard, Sebastian, 4
Saunders, W. B. and Co., 24*a*
Sydenham Society, 48

Thomas, Charles C., 25*a*, 25*b*
T[homas], E[dward], 49

Urie, R., 15

Wolfgang, Abraham, 41

Young, James, 43

GENERAL INDEX

(The numbers refer to the pages of the book)

For EU product safety concerns, contact us at Calle de José Abascal, 56–1°,
28003 Madrid, Spain or eugpsr@cambridge.org.

www.ingramcontent.com/pod-product-compliance
Ingram Content Group UK Ltd.
Pitfield, Milton Keynes, MK11 3LW, UK
UKHW060313090126
466816UK00021B/476